THE DESIGN & DRAFTING of PRINTED CIRCUITS

REVISED EDITION

by Darryl Lindsey

founder of the renowned MASTERS P.C. DESIGN SCHOOL

Published by

The Innovators
Bishop Graphics, Inc.
· Westlake Village, California 91359

Distributed by

McGraw-Hill Book Company

*New York St. Louis San Francisco Auckland Bogotá
Düsseldorf Johannesburg London Madrid Mexico
Montreal New Delhi Panama Paris São Paulo
Singapore Sydney Tokyo Toronto*

Distributed by

McGraw-Hill Book Company

*New York   St. Louis   San Francisco   Auckland   Bogotá
Düsseldorf   Johannesburg   London   Madrid   Mexico
Montreal   New Delhi   Panama   Paris   São Paulo
Singapore   Sydney   Tokyo   Toronto*

The Innovators
Bishop Graphics, Inc.

5388 Sterling Center Drive
P.O. Box 5007
Westlake Village, California 91359

Printed in the United States of America

First Printing, 1979

McGraw-Hill ISBN No. 0-07-037844-4

PREFACE

This second edition of The Design and Drafting of Printed Circuits adds over a year's worth of accumulated enhancements and updates to the original publication, which enjoyed four substantial printings. Now more than ever, the USE, HOW, and WHY of printed circuit design are thoroughly explained and illustrated. The second edition of this book is a teaching tool for students of the art and much more. It has been rounded out, expanded and fine tuned to cover a broader scope of readership, from students to industry professionals.

As a teaching tool, this textbook guides the student into analyzing problems by explaining the art's guiding principles, defining its terms and outlining design and manufacturing processes. As a reference book for industry professionals, it thoroughly explains advanced concepts such as multilayer boards, flexible circuits and the techniques of CAD, EMI suppression and automatic assembly by robots.

Almost half of the book has been revised or expanded:

Chapter One now includes a detailed primer on electronic components and schematic diagrams. Chapter Eight now includes inside-track information on multilevel pc boards, their rationale, tradeoffs, layouts and registration techniques. Chapter Eleven has been completely revised, expanded and integrated with a sequence of real-world photographs that illustrate the exacting step-by-step process of Artmaster Tapeup.

Chapter Five on Logic provides a source of information that is difficult to find elsewhere. The understanding of logic as it relates to pc design greatly increases the value of this book as a text and as an important source of general information. Chapter Seventeen on flexible printed circuits is entirely new, and so is Chapter Eighteen, a detailed account of how to shield electronics against EMI—the everpresent "soup" of electro-magnetic interference in the earth's atmosphere.

Chapter Nineteen—also newly added to the Second Edition—explains computer-aided design (CAD) systems and how to use them, and Chapter Twenty adds a checklist and overview of the entire book. A checklist has been included following Chapter Fifteen. This checklist is an aid to the inspection of printed circuit board master drawings.

The author's vast experience and state-of-the art materials are presented in a manner designed to help students and professionals in the solution of printed circuit design problems—to readers of this second edition it is the author's fond wish that this book will prove to be of lasting value.

ACKNOWLEDGMENT

The author would like to express his thanks for all the assistance that has been provided for the various editions of this book. I am much indebted to former students, colleagues and friends for their time and efforts in helping prepare this book. I am indebted especially to the following: Bill Balingit and Tom Walker.

And finally to my wife, Priscilla Lindsey for her time and patience in helping me prepare the manuscript.

INTRODUCTION

Evolution of the printed circuit board does not "just happen". In fact, some rather serious thought must be given to each individual project before actual design work can begin. Once started, the design should proceed through six crucial phases. These phases are outlined in this introduction and are explained in detail within the corresponding section of this book.

Pre-design effort will ensure that the printed circuit board designer has a general knowledge of the project ahead of him, electrically and physically. Most of all, it will ensure that his ideas and those of the circuit design engineer coincide. Coordination between the circuit design engineer and the PCB designer *cannot* be overstressed. This communication link is of the utmost importance if the end product is going to meet *all* fit, form, and function requirements set forth.

Before work of any kind can be started, however, the circuit design engineer must prepare an appropriate "engineering" schematic. The engineering schematic is transferred to the PCB designer, and phase 1 of her/his work begins at this point. All too often, the schematic phase is not approached "head-on" and the remaining design phases suffer accordingly. To insure that you get off on the right foot, put that engineering schematic at the *top* of your "list of tools".

Printed circuitry and the printed circuit board, what are they? For years, printed circuitry and printed circuit boards have been defined and re-defined within the electronics industry. The following are definitions that have been extracted from leading electronics dictionaries:

> "A circuit in which the interconnecting wires have been replaced by conducting strips printed, etched, etc. onto an insulating board. It may also include similarly formed components on the base board."

> "Also called a card, chassis or plate. An insulating board onto which a circuit has been printed."

Of course, these definitions assume that the reader has had prior experience with point-to-point wire soldering. For those unfamiliar with the process, it simply required hand-soldering insulated wires between component leads where electrical connection was required. Simple it was. However, it was also time consuming and the end product was complicated, bulky, and truly a production nightmare. Understandably, compactness, efficiency in assembly, and standardization have since become the goal of electronic industries in the United States and abroad.

In contrast to modern electronic components used today, the "printed wiring board" (as it was formerly known) included designs with bulky, high-heat producing components such as vacuum tubes, large oil-filled capacitors and transformers. Miniaturization and micro-miniaturization seen in today's electronic components (such as diodes, transistors, integrated circuits, capacitors, resistors, and transformers) have played a leading role in speeding up and developing the techniques and standards employed in the production of today's printed circuit boards. This book is dedicated to *these* techniques and standards.

The author prefers to define the modern-day printed circuit board as follows:

"The replacement of hand-soldered, point-to-point wire connections with thin lines of copper. These copper lines are affixed on one or both sides of flat, rigid, glass-epoxy insulating boards through various processes including photography and chemical etching. The board facilitates the rapid assembly of active, passive, discrete, non-discrete, and hybrid electronic components with the following result: A single compact assembly where ease of assembly, maintenance and reliability are an order of magnitude better than ever before possible."

TABLE OF CONTENTS

Chapter

1

SCHEMATICS

SCHEMATIC DIAGRAMS

INTRODUCTION

A schematic diagram consists of a system of graphic symbols that represent electronic, electrical, and electromechanical components (figure 1-1). The components are connected electrically as shown by the interconnecting lines. The schematic is the first step in a new or revised electronic design because it displays and identifies the components that make up the equipment. Your principal concern as a PC Designer is the conversion of the schematic diagram to an artmaster; therefore, learning to read a schematic is of prime importance.

This section contains the descriptions of all the component parts of the schematic diagram in figure 1-1. This diagram demonstrates the basic parts of a schematic. Each symbol, its reference designation, its fundamental purpose, and an introduction to the basic schematic layout, are all discussed in this section.

OBJECTIVES

After you study the material in this section, you should be familiar with all the essentials of the schematic in figure 1-1, and be able to:

1. Recognize the schematic symbols presented in this section.
2. Recognize the reference designations presented in this section.
3. Understand basic definitions of the initial electronic components in figure 1-1.
4. Know the fundamentals of schematic layout.
5. Define a schematic diagram.
6. Know the fundamentals of reading a schematic.

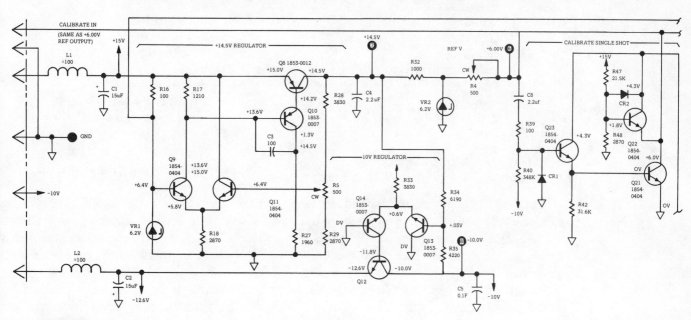

FIGURE 1-1
Example of
Schematic Diagram

SCHEMATIC SYMBOLS AND REFERENCE DESIGNATORS

The following is a discussion of each circuit component and symbol on the schematic in figure 1-1. A brief description of their purpose, application, and electrical operation is also given. An indepth theory of operation for each part is impractical since you, as a PC Designer, are not required to know electronic theory.

TABLE 1-1   REFERENCE DESIGNATIONS

A	assembly	MG	motor-generator
AT	attenuator; isolator termination	MK	microphone
		MP	miscellaneous mechanical part
B	fan; motor		
BT	battery	P	electrical connector (movable portion); plug
C	capacitor		
CP	coupler		
CR	diode; diode thyristor: varactor	PS	power supply
		Q	transistor: SCR: triode thyristor
DC	directional coupler		
DL	delay line	R	resistor
DS	annunciator; signaling device (audible or visual); lamp; LED	RT	thermistor
		S	switch
		T	transformer
		TB	terminal borad
E	miscellaneous electrical part	TC	thermocouple
		TP	test point
F	fuse	U	integrated circuit; microcircuit
FL	filter		
G	generator	V	electron tube
H	hardware	VR	voltage regulator; breakdown diode
HY	circulator		
J	electrical connector (stationary portion); jack	W	cable transmission
		X	socket
		Y	crystal unit (piezo-electric or quartz)
K	relay		
L	Coil; inductor	Z	tuned cavity; tuned circuit
LS	meter		

SCHEMATIC DIAGRAM REFERENCE DESIGNATORS

Each component is assigned an alphanumeric reference designation such as C1, R1, S1, Q1, etc. The letter (or alpha) characters C, R, S, Q, etc., identify the component: C for capacitor, R for resistor, etc. The number differentiates between electrical components of the same type: C1, C2, C3, etc. The use of reference designators provides an easily recognizable identification for every component part on the schematic. The letter that identifies the component is generally the first letter of its name or function. A for Assembly, C for Capacitor, CB for Circuit Breaker, CR for Diode (but original function was Crystal Rectifier), R for Resistor, and so forth. Since there are many names that begin with the same letter like tube, transistor, and transformer, it is not always practical to use the first letter; however, in most cases the letter gives some indication as to the name of the component. Table 1-1 provides a list of the most common reference designations.

SCHEMATIC DIAGRAM SYMBOLS

To simplify schematic diagrams, symbols are used to represent the various circuit components. In most cases, the symbol is representative of either what the component actually does in the circuit or how it is physically constructed. For example, lead-acid batteries are constructed of interwoven plates (figure 1-2.). Schematics use symbols.

The schematic symbol for a battery is a series of equal spaced short lines (representing the negative terminal) and long lines (representing the positive terminal) Figure 1-3.).

The similarities between the way a battery is constructed and its symbol is readily apparent. Another example is a capacitor. Some of the first capacitors were constructed of two metal foils separated by an insulator. This was rolled to make a cylinder. A capacitor can also be charged and store electricity similar to a battery. Therefore, the symbol simply became two parallel plates of equal length and separated by a given space (figure 1-4.).

In almost every case, electronic symbols evolved logically from either their circuit application, their construction, or from a combination of both.

SCHEMATIC DIAGRAM COMPONENT DESCRIPTIONS

The descriptions of the components will cover their basic purpose in the circuit, and their fundamental operation and construction. Refer to Figure 1-5 for the symbols, explanation of the symbols, and reference designation for each component.

CAPACITOR

The capacitor is also known as a "condenser" which name is derived from the part in an automobile's distributor. A capacitor is used to (1) bypass (short) noise and unwanted ac signals to ground (bypass capacitor), (2) filter dc voltages in power supply application (filter capacitor), and

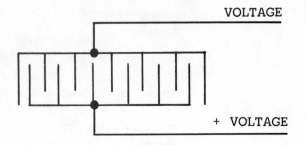

FIGURE 1-2
Lead-Acid Battery
Construction.

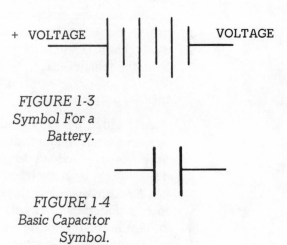

FIGURE 1-3
Symbol For a
Battery.

FIGURE 1-4
Basic Capacitor
Symbol.

(3) couple or transfer signals from one circuit or stage to another (coupling capacitor). There are also capacitors connected to a dc bus that appear to be filter capacitors but they prevent interaction between circuits and are called decoupling capacitors. A capacitor stores electrical energy, blocks the flow of direct current (dc), but permits the flow of alternating current (ac). The ability to store electrical energy is called capacitance; the unit that measures capacitance is the farad. (Farads are extremely large units so the more common units are microfarads.) A capacitor consists essentially of two conducting surfaces separated by an insulating material or dielectric such as air, paper, mica, glass, plastic, film, or oil.

CONNECTORS

The PC Board in Figure 8-4 shows the connector tongue or edge connector through which all input and output signals, voltages and grounds are routed. The connectors of a PC Board are usually pin contacts and are "plugged-into" a socket or receptacle. The male or plug connector symbol is shown in figure 1-5B (1). The female socket (jack) or receptacle symbol is shown in figure 1-5B (2). The plug or male contacts are always the removable part of a connector assembly as shown in figure 1–5B (3), and the receptacle or jack is the stationary or fixed part. The male or pin contacts are not wired to voltage or power sources. This practice is to protect personnel from touching the pins (a shock hazard) when the plug is separated from the female jack.

DIODE

Also called a crystal diode, crystal rectifier, and semiconductor diode. Its principal purposes in a circuit are as a switching device and a detector (rectifier). A rectifier is a device that allows current to be passed in one direction (forward) but blocks the current in the reverse direction.

Diodes are constructed from semiconductor material*. Impurities are added to determine the types of conductive properties: either n-type or p-type. (See figure 1-5J discussion on transistors.) The name diode means it is a two-electrode device: one electrode is made of n-type semiconductor materials and the other is p-type. The junction of the two dissimilar materials produces the properties of a rectifier.

INDICATOR, SIGNALING DEVICE

When you switch most radios or stereos to ON, a lamp illuminates. The lamp indicates the condition that power has been applied to the radio or stereo. The lamp is

*Semiconductor materials include germanium, selenium, and silicon and have resistive properties halfway between an insulator and a conductor.

then a signaling device, signaling that the radio is ON or should be. Any signaling device, audible or visual, is in this category (figure 1-5D). The audible signals are such devices as alarms and buzzers; the visual devices refer to illuminating indicators such as pilot lights, indicator lamps, LEDs (light emitting diodes), and such. (Gauges, clocks, meters, and timers are not in this category.)

INDUCTOR

The inductor is also known as a coil since it is constructed by winding a conductor into a coil. An inductor provides opposition (impedance) to alternating or pulsating current in a conductor, but offers very little impedance to direct current (dc). The electrical value of a coil is called inductance and is measured in Henries. Inductance is developed in a conductor by the motion of the conductor in a magnetic field or the motion of the magnetic field across the conductor. An air core inductor (Figure 1-5. E) is constructed by making loops or coils of wire around a form, which is later removed, or around an insulator such as a plastic or ceramic support. An iron core inductor is constructed by making coils of wire around an iron core or other magnetic material.

RELAY

A relay is an electromechanical device in which either normally-open relay contacts are closed or the normally-closed contacts are opened. When a relay is idle in a circuit, it is said to be in a "deenergized state." (All relays in figure 1-5F are shown deenergized.) An activated relay is energized. A relay, then, is a control. It uses the energizing of a magnetic coil (solenoid) in one circuit to control the opening or closing of contacts in a completely different circuit. Therefore, the relay solenoid and the contacts of that relay may appear in different locations on the diagram.

SCHEMATICS

RESISTOR

Resistors (1) control or limit the amount of current flowing in a circuit and (2) provide a voltage drop. The composition of a resistor, the material it is made of and the density of that material, produces a specific amount of resistance or opposition to current flow. This resistance is measured in ohms; such as 100 ohms, 2000 ohms, etc. Ohms is also expressed by the symbol " Ω " which is the Greek letter Omega. Resistance would then be written: 100Ω, $2,000\ \Omega$, etc. Resistors are made of carbon, metal film, and wire-wound. Since it is impossible to manufacture a perfect 100 or 200 ohm resistor, the manufacturer is allowed some percentage of error. If a plus-or-minus 10% (±10%) error is allowed for a 200 ohm resistor, then the resistance value may vary 10% or 20 ohms either higher (+) or lower (−) than 200 ohms. Its value may be as much as 220 ohms or as little as 180 ohms. Precision resistors of ± 1% and 2% tolerance achieve greater accuracy but with higher cost.

Resistors are also rated according to the amount of heat they can dissipate; this is called wattage rating. Any opposition to current flow in a circuit causes heat. (Think of the coils in a toaster or heating pad which are special kinds of resistive elements.) A resistor must tolerate this heat without burning up. If a circuit requires excessive current (current and wattage have a direct relationship) then the resistors used must have a higher wattage rating. Resistors are normally ¼ Watt (¼W) and ½ Watt (½W). Digital circuits of today have low current requirements and can use 1/8 Watt resistors.

SWITCH

A switch is a device that connects, disconnects, or transfers one or more functions or circuits to one or more destinations. The common mechanical switch, prevalent in all homes either as an on-off switch or selector switch with two, three, or more positions, still has extensive applications in electronics (figure 1-5. H). There are also electromechanical switches and electronic switches. The electromechanical switches perform like relays. The electronic switches come as microcircuits which the PC Designer treats the same as other microcircuits.

TRANSFORMER

The word transformer is derived from the word "transform" which means to convert something from one magnitude to another, or from one type to another. Therefore, a transformer is an electrical device which, by electromagnetic induction, transforms electric energy from one or more circuits to one or more other circuits at the same frequency, but usually at a different voltage and current. A transformer is constructed so one winding (primary) induces voltage into a second winding or windings (secondary). When a transformer transfers electric energy at a different level it is called either a step-up or a step-down transformer. If there is no level change, then it is said to have a 1:1 voltage ratio. A step-up transformer increases the voltage as energy is transferred from the primary to the secondary winding. A step-down transformer decreases the voltage as energy is transferred from the primary to the secondary.

TRANSISTOR

A semiconductor device has electronic characteristics which enable it to control voltage and current gain. In a circuit containing resistors, capacitors, diodes, and transistors, the transistor is the central device, the active component in the circuit. Transistors provide amplification, rectification, switching, detection, and almost any function in an electronic circuit.

VOLTAGE REGULATOR OR BREAKDOWN DIODE

A voltage regulator provides a nearly constant voltage at its output, regardless of normal input-voltage variations and virtually independent of the current. Voltage regulation is possible in certain semiconductor diodes because of a characteristic called avalanche breakdown. Normally a breakdown condition is associated with trouble or nonoperation. In this case, however, avalanche breakdown is a nondestructive or controlled breakdown. It is used in many applications in electronics: as a voltage regulator, voltage limiter, and as a voltage reference diode. (It is used frequently in present day circuits as a voltage reference.)

SYMBOLS

The electrolytic capacitor requires that the more positive voltage potential be connected to the straight line of the symbol which is always marked with a plus (+).

The symbol for a capacitor (1) is derived from the basic construction of the capacitor which has two metallic surfaces, separated by an insulator or dielectric. The capacitor symbol (2) is as widely used as (1) and is gaining acceptance. It is drawn with one plate as a straight line and the other curved. This symbol evolved from the symbol for an electrolytic capacitor in (3).

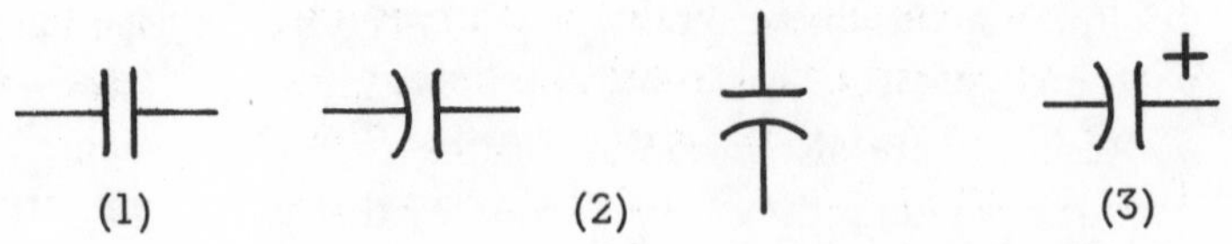

FIGURE 1-5A
CAPACITOR
Reference Designation: C

(1) (2) (3)

The symbol for a male contact is not the head of an arrow and the symbol for a female contact is not the tail of an arrow. Three elements are required to be shown on a schematic diagram at each contact: (1) the type of contact shown graphically as either a plug (P) or a jack (J); (2) the connector reference designator P1, P2 or J1, J2, etc.; and (3) the contact Pin letter or number.

NOTE

The connector contacts or terminals are numbered on the diagram as it is convenient to the location of circuits and not in alphabetical or numerical sequence (4). See Figure 1-5B.

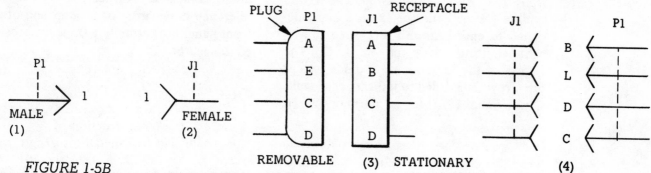

FIGURE 1-5B
CONNECTORS
Reference Designations: P and J

Since the diode is a two-element (two-electrode) device, its symbol (1) shows the two electrodes: the cathode and the anode. These terms are adopted from its vacuum-tube counterpart which was also called a diode with an anode (plate) and a cathode. (The arrowhead of the symbol points in the direction of *"conventional"* current flow. The *"electron flow"* is still from the cathode to the anode as in the diode vacuum tube.) To help you remember which electrode is the cathode: in (2), the cathode forms the letter "K" (for Kathode) when the anode is separated.

NOTE

You may occasionally see the reference designator "D" used for diode but this is seldom used since CR for crystal rectifier has the widest usage and is also the military standard.

FIGURE 1-5C
DIODE
Reference Designation: CR

The symbol for an indicating lamp (1) is derived from the construction of the filament of an incandescent lamp. The second symbol for an indicator (2) is a diode designed to emit light and is referred to as a light emitting diode or an LED. The symbol is the same as for a diode except the arrows are added to indicate light emission. Symbol (3) is for an alarm or buzzer.

Some companies still use an "I" as the reference designator for a lamp and other front-panel indicators but DS is the accepted designator.

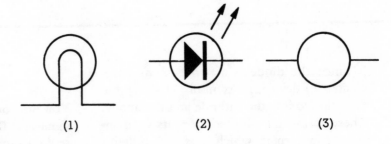

FIGURE 1-5D
INDICATOR, SIGNALING DEVICE
Reference Designation: DS

(1) (2) (3)

The symbols for an inductor or coil are easily recognized since they graphically attempt to show the construction of the inductor. The symbols in (1) depict a core or center of the winding as air. It is constructed by wrapping the conductor around a form and then removing the form. Also, an air core inductor may be constructed by winding the conductor around a permanent, non-magnetic, insulating material (rigid form) such as cardboard, plastic, or ceramic which remains as part of the coil. The second symbol of (1) drawn with open loops is less frequently used because it is difficult to draw. In (2), the two straight lines indicate that the coil is wound around an iron core or some other magnetic material. The iron core increases its inductive intensity.

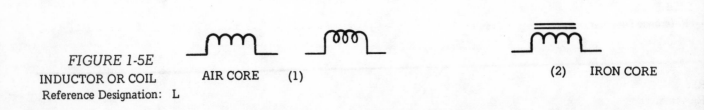

FIGURE 1-5E
INDUCTOR OR COIL
Reference Designation: L

AIR CORE   (1)

(2)   IRON CORE

The symbol for a relay resembles its actual construction and circuit operation. The relay in (1) contains two normally open (N/O) contacts and one normally closed (N/C) contact. The contacts of relay (2) are normally closed. In a schematic presentation, a relay is usually shown de-energized. When the relay in (1) energizes, the normally open (N/O) contacts close and the normally closed (N/C) contact opens. Another symbol for the relay is shown in (3). Note that the winding of the relay (solenoid) is like the symbol for an inductor. It is common, then, to call the solenoid a coil.

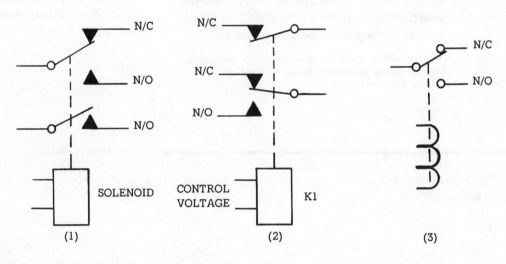

FIGURE 1-5F
RELAY
Reference Designation: K

The symbol for a resistor (1) is a series of peaks and valleys which makes a zigzag. This symbology is an attempt to show an increased amount of material or a harder or longer trip for the electrical current which would depict the property of resistance. This is the standard symbol and the most commonly used symbol for a resistor. It is also the symbol used to show the electrical property of resistance or total impendance in an electronic device or circuit. Although it may not be apparent, there are exact drafting standards for drawing this resistor symbol. The symbol must have three peaks and three valleys drawn at 60 degrees from the horizontal. However, the symbol is easily drawn freehand and engineers and technicians will jot it down with a couple of fast squiggles. The second symbol (2) has little use on schematics and only limited use in certain industrial circuit designs. However, it is used exclusively in PC Designing. The size of the rectangle indicating the exact physical size of the resistor on the PC layout. More information on resistors will be given in later sections.

FIGURE 1-5G
RESISTOR
Reference Designation: R
(1)   (2)

A standard symbol for a switch (1) is the single-pole, single-throw (SPST) switch* If this represented an ON-OFF switch, the schematic would show S1 in the OFF position or normally open (N/O) position. The symbol (2) shows a single-pole, double-throw (SPDT) switch. If this was an AM/FM selector switch, then the AM and FM positions of the switch must be shown on the schematic. If one position is selected on a switch (FM in this example), then all other selectors must also be in FM. Switch positions must show a consistency and continuity throughout the schematic and also

*Toggle Switch. Common name for simple two-position switch whose movement through a short arc closes or opens electrical circuits.

switch representation (3) is used in industrial applications. This symbol is used to show both switch and relay contacts. The symbol is for the SPDT switch (2). The slanted line across the two parallel lines indicates contact is made.

NOTE

The symbol in (3) if it represents a relay, is shown deenergized. So FM is the normally closed (N/C) contacts and AM is normally open (N/O) contacts.

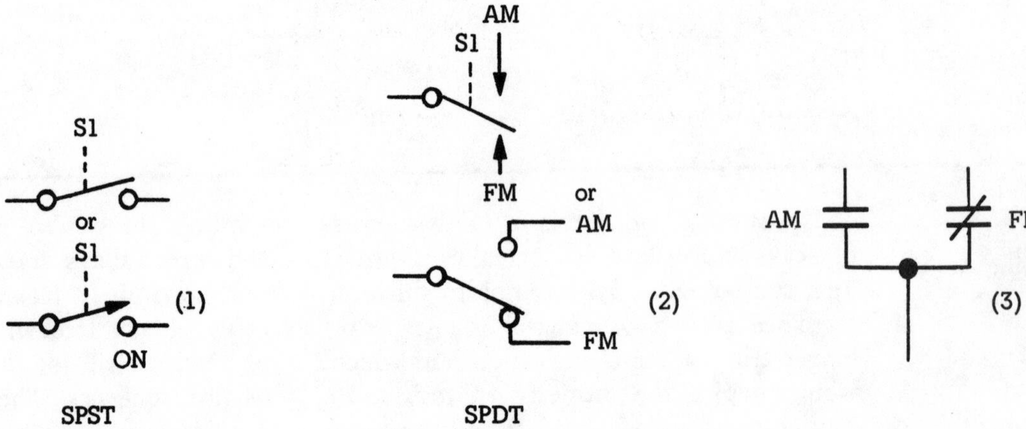

FIGURE 1-5H
SWITCH
Reference Designation: S

The transformer symbol is derived from the construction of the transformer which consists of two or more windings wound on a common core. The windings are positioned adjacent to each other to obtain magnetic or inductive coupling. The windings of a transformer resemble those of inductors because the same electric principle is involved in transformers and inductors. The input voltage is applied to the primary (PRI) winding and the output is taken from the secondary (SEC) winding. Almost always the primary is on the left so the signal flows left-to-right with the output at the right. The core or center (1) is air or a non-magnetic, insulating material such as paper, plastic, or ceramic around which the primary and secondary conductors are wound. The air-core transformer is used for coupling signals between stages at higher frequencies. The next transformer symbol (2) has an iron core which increases the inductive coupling. The power transformer (3) usually has two or more secondary windings. These windings are associated with the power supplies and the generation of different voltage levels required in sophisticated circuits. The center tap (CT) is a neutral leg, available for certain power circuits which will be discussed in later sections.

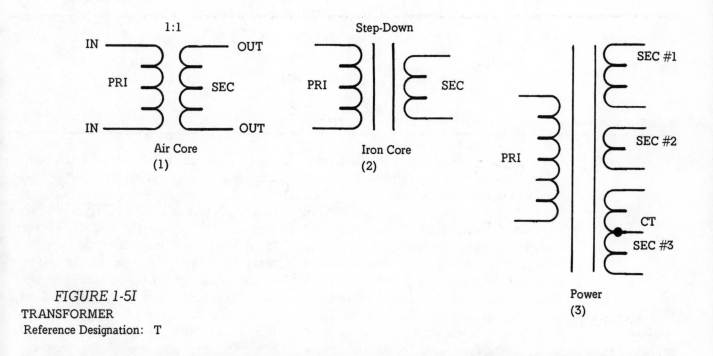

FIGURE 1-5I
TRANSFORMER
Reference Designation: T

The simple transistor used as an amplifier requires three connections. Since transistors evolved when electron tubes were *the* active device, the transistor symbol was patterned after a triode vacuum tube (3) which also has three connections. An input control element (base/grid), a source of electrons (emitter/cathode), and an output element (collector/plate). The arrowhead was added to distinguish between the emitter and the collector, and to show the direction of "conventional" current flow. Electron flow is opposite the direction the arrow points.

A transistor is formed by two junctions of dissimilar semiconductor materials called "N" and "P" types* (Figure 2-6J (4)). The semiconductor materials most often used are germanium (Ge) and silicon (Si). Because the transistor is formed by two junctions, the numbering system for transistors is prefixed by a "2" or 2N1132, 2N1893, 2N3440, etc. Whereas, a semiconductor diode is formed by one junction and its numbering system is prefixed by a "1" or 1N21, 1N43A, 1N645, etc. A semiconductor with three PN junctions would be 3N---. The numbers following the 1N or 2N, in all cases, do not indicate the type, function or use of the device. Such information is found on manufacturer's data sheets or in data books.

The circle around the three elements may not always be shown. Most engineers omit the circle and draw only the elements. However, the circle is always drawn on final schematics or illustrations.

An aid in remembering the type of transistor: check the direction the arrowhead is pointing. If it is *Pointing iN* (1), it is a PNP; if it is *Not Pointing iN* (2), it is an NPN.

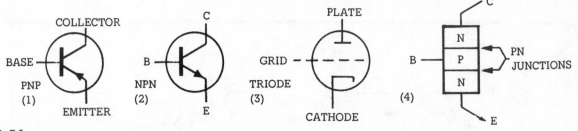

FIGURE 1-5J
TRANSISTOR
Reference Designation: Q

The symbol contains a diode in a circle (1) with another graphic symbol. The other graphic symbol defines it as a breakdown diode. (Other graphic symbols are used to denote other types of diodes which are explained later in the book.) The reference designator is VR which was derived from a vacuum tube commonly called a VR tube. With the semiconductor diode, other reference designators have also been used and since you will encounter these in your work, a few comments follow concerning them.

Originally the semiconductor voltage-controlling diode was called a zener diode** and "Z" was used as the reference designator. (The name "zener diode" is still the most common name for the avalanche breakdown diode.) The reference designation "CR" also applies to the breakdown diode when it is used as a rectifier; however, it is still acceptable to assign VR even in this application. Also, the reference designation "D" (for diode) is still used.

The avalanche breakdown diode is called by several names, has many uses, and may have different reference designations, but this is no problem for the PC Designer. On the schematic diagram, all engineers indicate the diode by either symbol (1) or (2). The reference designator selected by the engineer depends on the application: VR for voltage regulators, voltage reference diodes, or limiters, and CR for rectifiers. You, as a PC Designer, in your schematic layout, will select the reference designator and symbol accepted and set by your company standards.

*N-type material has an excess of negatively charged particles called electrons; P-type has a deficiency of electrons.
**Originally the avalanche breakdown diode was called a Zener diode. This happened before it was learned that the Zener effect had no significant application in the operation of diodes of this type.

14   SCHEMATICS

To aid in distinguishing this diode from a regular "CR" diode, the engineers almost always label the breakdown diode with its specified breakdown (voltage rating). Since this voltage is of value to the engineer in the circuit design, the value is generally noted near the diode. For example, some common values are 6.8V, 7.2V, 9.6V, etc. The engineer will normally draw a wavy line at the cathode end of the diode similar to that shown in symbol (2).

FIGURE 1-5K
VOLTAGE REGULATOR OR BREAKDOWN DIODE (1) (2)
Reference designation: VR

REFERENCE DESIGNATORS IN PARTS LISTS

The use of reference designators in a parts list or material lists makes it possible for the list to be organized in an alpha-numerical sequence which aids in listing and locating parts quickly. The parts list has the complete reference designator listed by assembly and component number. If C1 were on assembly number one, then the full reference designator would be A1C1. In present-day electronic equipment, an assembly is the same as a PC Board. It is common usage to refer to assembly one as board one, assembly two as board two, and so forth. Therefore, if component R14 is on assembly six (PC Board six), in the parts list you would first locate the A6 assembly (which follows A5), then go through the alphabetical sequence to the R's and then through the numbers to R14.

READING SCHEMATIC DIAGRAMS

The schematic diagram is the basic tool for the PC Designer and it is vital that you learn to read one quickly and accurately. The schematic diagram is therefore usually the first step in the design of electronic equipment or a PC Board, because it presents and lists all components that are to be included in the new design. The information on and associated with the schematic later becomes the source of identification and rating of the various components for design layout, purchasing, manufacturing, testing, service manuals, and field maintenance. A schematic diagram may contain any or all of the following elements: (1) electrical or electronic assemblies, subassemblies, controls, components, and mechanical parts shown as graphic symbols; (2) component identification with reference designations, component values, and part numbers; and (3) conductor paths showing circuit and component interconnections and signal flow. The schematic diagram is laid out according to electronic drafting procedures to produce an easy-to-follow signal flow with few cross-over lines, one that adheres to functional convention, and has readability.

SCHEMATIC FORMAT

The components, in the form of graphic symbols, are arranged in groups or individual circuits according to their function. The groups are called stages or simply, circuits. The symbols for passive components such as resistors, capacitors, inductors, etc., are grouped functionally around the symbols of the active components such as tubes, transistors, integrated circuits (ICs), etc. Each such grouping performs a particular function and is called a "stage" of operation. These groups of components or stages are drawn on the schematic in established ways that are according to set practices, methods and standards. (These standards are discussed in detail in later sections of this book.) It is part of the PC Designer's job to become thoroughly acquainted with these drafting standards. This standardization has been developed over the years by those who use schematic diagrams such as engineers, technicians, service and maintenance personnel, drafters and others.

The normal signal flow on the schematic is from left to right and from top to bottom, with the inputs on the left and the outputs on the right. The signal flows into a stage, the stage performs its electronic function on the signal, and then the signal is passed (coupled) onto the next stage. Such a stage or function in your radio, for example, is the power amplifier. As the name suggests, the power amplifier stage amplifies the signal to increase the power. Then it is coupled out to drive a transducer such as speakers or ear phones.

Secondary circuits (usually non-signal or auxiliary circuits) such as power supplies, filters and control circuits may be grouped at the bottom of a diagram. In some schematic configurations, it is common for the main circuit element on a PC Board to be centralized on the schematic with inputs entering on the left and the outputs exiting on the right.

NOTE

The position of the circuit on the schematic does not determine its importance. The circuit stages across the top may have no greater importance than those in the center or towards the lower part of the diagram. The circuits are arranged on the diagram according to best drafting practices and standards. The drafter lays out the diagram to alleviate circuit and component congestion and to make best use of the space available, to provide a natural, easy signal flow, and to group the circuits functionally—but not according to their importance.

All signals entering or leaving the PC Board and all voltages applied to the board are routed through connectors. The connectors on the PCB are called plugs and they mate with external connectors called jacks. It is common practice, as mentioned above, to have the output connectors on the right. However, this is not always possible and some outputs may be on the left as well as some inputs entering from the right. You may not be able to distinguish between an input and an output on your first look at a schematic, but with time it will become apparent to you.

COMPONENT PLACEMENT

The placement of the components on the schematic does not represent the actual physical location of the component on the PCB nor does it represent the physical size. (However, in some cases, information on the schematic may indicate the physical size, as would the wattage rating of resistors.) The component's position on the schematic, whether it be shown vertically, horizontally, or at any angle, does not affect its function or how it is to be connected on the PC Board. The PC Designer only needs to ensure that all components that are connected to a common line on the schematic are connected together on the PC layout.

A line on the schematic should be regarded as a point; therefore, a line is merely an extension of a point. Two components connected to the same line but separated on the schematic by 6 or 7 inches must be viewed as being connected to the same point. In designing the PC Board, these two components must be regarded as being electrically connected together and the long-line separation is nonexistent. Furthermore, if five or six or seven parts are connected to the same line (point), the order or arrangement on the PC Board, would not be affected.

POWER SUPPLY AND GROUNDS

Each individual voltage and ground connection is not physically tied to a common line on the schematic. (This would add many unnecessary lines.) Instead, voltage and ground symbols are used to eliminate voltage and ground connection lines. The voltage symbol is an arrow (figure 1-6) at which point is also indicated the voltage value. There may be as many as four or five different voltage values on a diagram and each would show the voltage symbol and voltage value. (At the connector would be shown the pin on which each voltage enters the board, along with the voltage symbol and voltage value.) The standard ground symbols are shown in Figure 1-6B. Other information on grounds is given later in the book.

Figure 1-7 demonstrates how the voltage and ground symbols are used. On the schematic they are not shown connected; however, on a PC Board all grounds are connected together, and all +15V are connected to a common point as well as the – 15V.

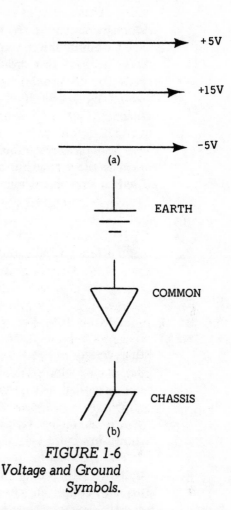

FIGURE 1-6
Voltage and Ground
Symbols.

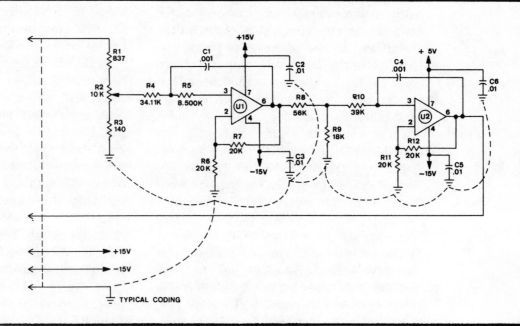

FIGURE 1-7
Interconnections
to a Ground
Point. Dashed
lines are not
drawn in, they
merely show the
electrical
interconnection.

SCHEMATICS

When connections are made to a common line or point, the common point is referred to as a "bus" or a ground line, and would be called a "ground bus." The +15V line would be called the "+15V bus" and so forth.

The voltage symbol most commonly used is the arrow but any symbol may be used at the discretion of your company.

The Schematic diagram is composed of symbols that represent electrical and electromechanical components plus lines that represent conductors which interconnect components. The components are assigned reference designations to differentiate between similar symbols. The schematic of Figure 1-8 shows connecting lines, symbols and reference designations.

SCHEMATIC DEVELOPMENT

Generation of the schematic is a repetitive process that starts during the conceptual design phase and continues until the design is completed. The following lists many typical steps that occur during the schematic development cycle. A typical sketch and the resulting schematic are shown in Figure 1-8.

The schematic that the designer or drafter will receive from the engineer usually comes in the form of a rough sketch. This sketch should contain enough information to communicate what the engineer wants. The sketch often will be rough and on several sheets of paper. The designer must make sure that the sketch has all the symbols, component values, inputs, outputs, test points, reference notes and other necessary documentation, and is clear and accurate. Figure 1-9 shows a typical sketch for an analog circuit that a designer might receive from the engineer.

Once the rough sketch has been surveyed for any missing items, the designer must identify all components with reference designations. These designations consist of a combination of letters and numbers that identify both the symbol (and the component they represent) and their location on the schematic diagram. They are not, however, to be considered as an abbrevia-

tion for the names of the items. The letters used in the reference designation identify the class of the component, as shown in Table 1-1.

Reference designations are used not only on the schematic diagrams but also on the connection or wiring diagrams, parts lists, detail or assembly drawings, and technical manuals. These designations are the main link between electrical circuits and the mechanical details of electronic equipment. Without these reference designations it would be impossible to differentiate among identical components that appear throughout the schematic diagram.

Reference designations are classified in two general divisions—those used for non-controlled consumer products or simple components and those used for military and complex equipment such as computers. Even though each category appears to be at extremes to the other, there are many similarities. Therefore, it is recommended that the designer become well acquainted with a system that will meet both areas of reference designation. The most commonly used information can be found in The American National Standards Institute (ANSI) Standard Y32.16-1975, Reference Designations for Electrical and Electronics Parts and Equipment.

THIS SKETCH:

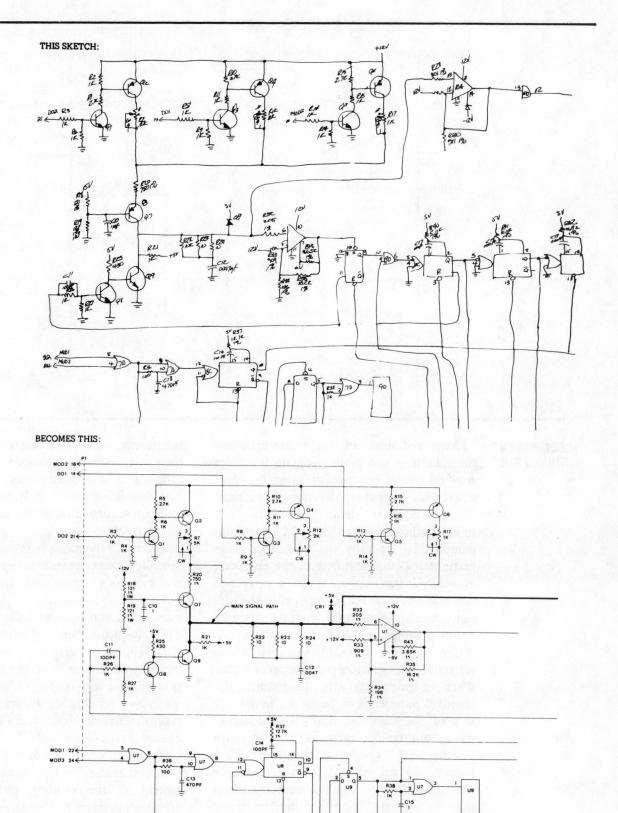

BECOMES THIS:

FIGURE 1-8

SCHEMATICS

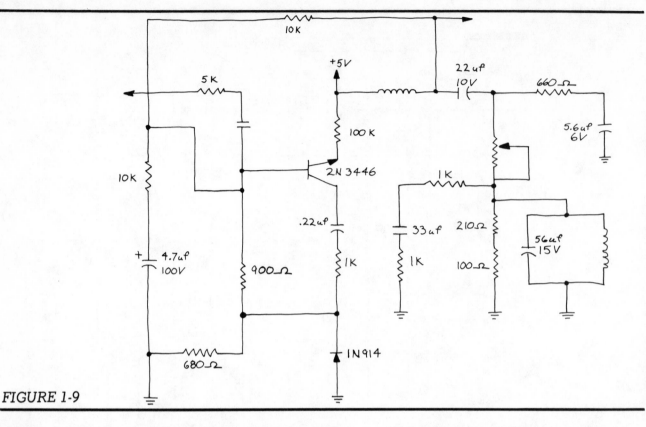

FIGURE 1-9

COMPONENT NUMBERING

After a designer or drafter has reviewed the sketch or schematic diagram from the engineer, each component must be identified with its proper reference designation. This procedure of identification is of paramount importance and should be carefully completed. Generally the sketch or schematic will be unidentified. If the engineer, in some instance, has assigned reference designations to each component, the designer must gain approval from the engineer before making any changes.

The designer typically will assign the reference designation to the schematic. When assigning reference designation, the standard practice is to begin at the top left side of the schematic and proceed downward, numbering each family of components with its own number in numerical sequence. Then proceed from left to right and top to bottom until all components in that family are identified by reference designation and number. Figure 1-10 illustrates this numbering sequence.

When a schematic diagram is drawn in a serpentine form or in two or more rows, the reference designation numbers are assigned as shown in Figure 1-11. The circuit flow is followed, applying the standard number sequence from top to bottom and left to right.

When a schematic is revised, adding or deleting components, it is not necessary for the reference designation numbers to be positioned consecutively. It is recommended to continue with the last reference designation number, and not add back into the schematic diagram those numbers which were deleted. A reference designation table is very helpful (when changes are made to a schematic). Figure 1-12 shows a typical reference designation table. This becomes particularly useful on very large schematics, when it would be difficult to locate the last used reference designation among all the symbols used. If careful attention is given to this table on the schematic, many errors will be eliminated in the numbering of the components.

GENERAL SCHEMATIC FLOW

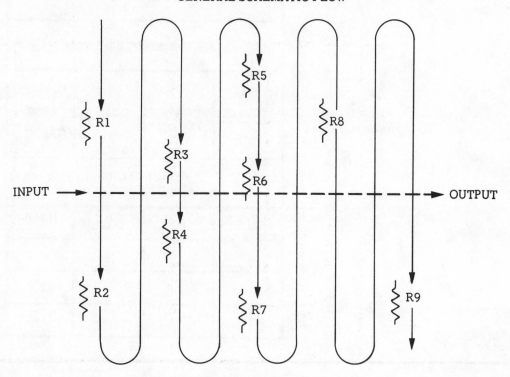

FIGURE 1-10

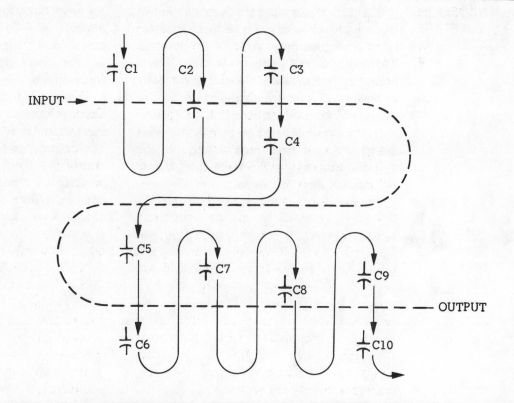

FIGURE 1-11

SCHEMATICS

REFERENCE DESIGNATION TABLE			
LAST DESIGNATION USED			
C54	L19	Q20	R39
DESIGNATIONS NOT USED			
C15	L9	Q2	R19
C19		Q19	R24
C36			R26
C51			

FIGURE 1-12

BASIC ELEMENTS OF SCHEMATICS

The schematic diagram is made up of three basic elements. These elements are: (1) the conductor paths, (2) the symbols and (3) the component identities. It is important to understand these three basic elements of the schematic in order to obtain the proper interpretation.

The conductor paths on most schematics are drawn with the inputs on the left side and the outputs on the right side. Figure 1-13 shows a typical signal flow. When the schematic diagram has more than one serpentine directional change, the inputs generally will be located on the upper left-hand side of the diagram and the outputs on the lower right-hand side of the diagram. Figure 1-14 illustrates this technique. In some instances, the inputs are located at the top left-hand side of the diagram, and the output at the bottom right-hand side. The outputs are preferred on the right-hand side of the diagram, although these outputs in certain instances will be located at the bottom. Figure 1-15 shows this technique.

SCHEMATICS

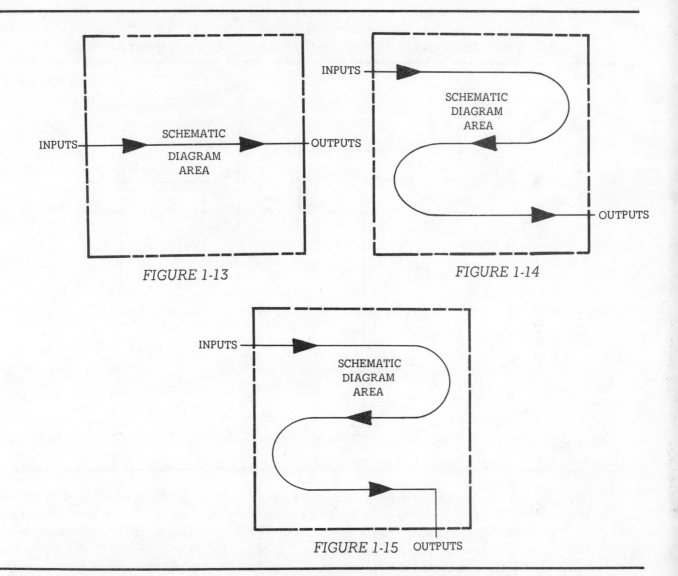

FIGURE 1-13

FIGURE 1-14

FIGURE 1-15

As mentioned previously, the normal signal flow on the schematic diagram is from left to right and from top to bottom, when the schematic diagram has one or more levels of stage circuits. Therefore, when the output of one stage flows through the coupling circuit and into the input of the next stage, this circuit arrangement is called a cascading circuit.

Many schematic diagrams have several voltages drawn on the diagram. These power signals are generally drawn as heavy lines, called bus lines, to differentiate them from signal paths. A good practice is to draw highest voltage on the uppermost line, with the others progressively decreasing to zero or minus voltages on the lower lines. This makes it easier to read the schematic diagram. A typical voltage sequence might be organized as follows:

+250	volts
+150	volts
+5.2	volts
0	volts
−5	volts
−15	volts

SCHEMATICS

23

In many instances, because of crowding, the voltages are terminated by using an arrow that shows the voltage level of the termination, instead of trying to carry the voltage to a common bus line. Figure 1-16 illustrates this practice.

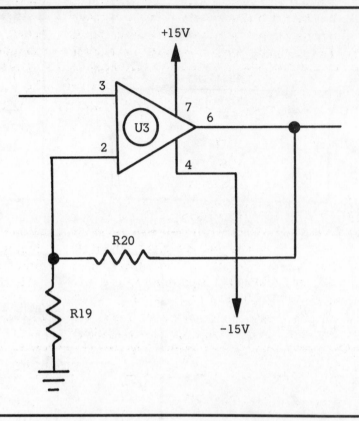

FIGURE 1-16

SYMBOL ARRANGEMENT One general practice in drawing a schematic diagram is to show the components in stages, with each stage centered around a transistor or IC, as illustrated in Figure 1-17. The symbols are drawn functionally around each stage, depending upon the signal path or current flow. This method of arranging symbols has become a standard for those who work with schematics on a technical level.

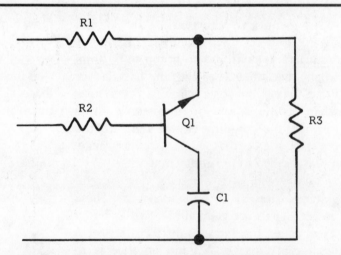

FIGURE 1-17

STAGE ARRANGEMENT

As the individual stage circuits are arranged for ease of connecting the signals together, another circuit, called a coupling circuit, is used to interconnect these stages. The coupling circuit consists of one or more components. The stage circuits are usually drawn in a horizontal direction with the power or auxiliary circuits drawn below the stages. Figure 1-18 gives an example of multiple stages connected.

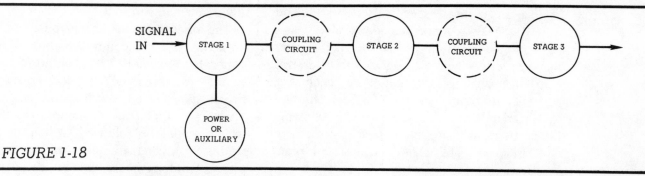

FIGURE 1-18

Understanding the coupling circuit is very useful when the P.C. designer divides the total schematic diagram into small and manageable circuits. The coupling circuit can be a resistor, capacitor or diode and therefore becomes an important consideration when deciding total area needed on either the schematic or PC board. Figure 1-19 shows a schematic divided and the coupling circuit connecting the two stages together.

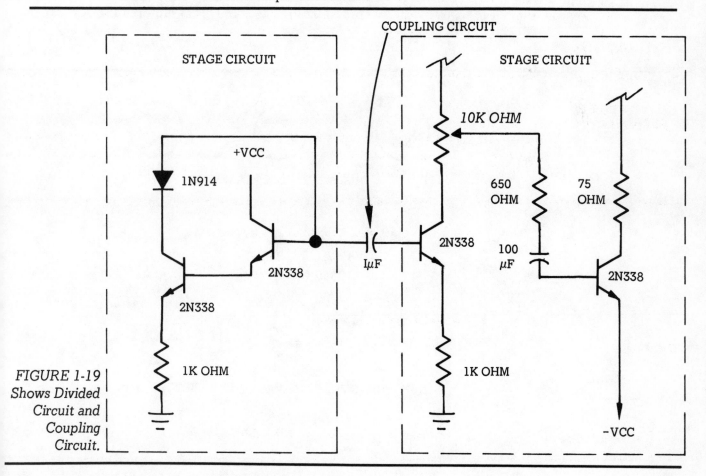

FIGURE 1-19 Shows Divided Circuit and Coupling Circuit.

In many companies today, the P.C. designer will be required not only to layout and tape the artmaster for a P.C. board, but also to draw a completed and formal schematic diagram of the P.C. board. Therefore, the designer must be well acquainted with the company standards, as well as commercial and military specifications (ANSI and MIL-STD), commercial data sheets and accepted drafting procedures before any successful schematic diagram can be drawn. The preceding information on schematics will also apply to the development and diagram layout and drawing. The following sections will give more insight into some basic schematic considerations and practices.

SYMBOL ARRANGEMENT

It is important to note that the orientation of a graphic symbol on the schematic diagram does not change the meaning of the symbol or its electrical function. Likewise, the placement of the symbol on the schematic diagram has no correlation to the actual physical location or arrangement of the component on a PC board or in an electronic system. The importance of the symbol or component relationship is that the symbol or component must be electrically connected correctly.

The orientation of the graphic symbol does not affect the interpretation of the symbol. Figure 1-20 illustrates some symbol arrangements. This does not apply to all symbols on a schematic diagram. Therefore, there are standard practices which should be followed to help in the reading of a schematic diagram. For example, ground symbols should be drawn pointing downward. (Figure 1-21A.) The voltage symbol should be drawn pointing upward for a plus (+) voltage, and downward for a negative (−) voltage. (Figure 1-21B.) Under crowded conditions it would be permissible to draw the ground and voltage symbols as shown in Figure 1-21C.

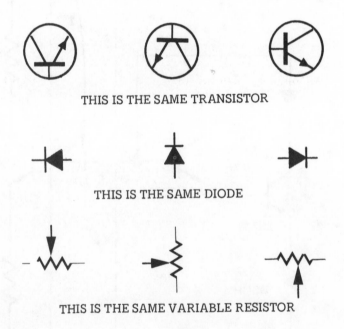

THIS IS THE SAME TRANSISTOR

THIS IS THE SAME DIODE

*FIGURE 1-20
Symbol
Orientation*

THIS IS THE SAME VARIABLE RESISTOR

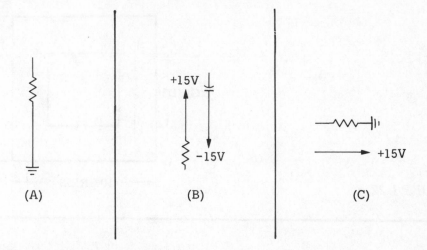

(A)   (B)   (C)

FIGURE 1-21

ESTIMATING SPACE

Before any attempt to draw a schematic diagram, the drafter must make two decisions. First, the finished sheet or format size must be established; second, the schematic must be drawn with all graphic symbols according to established standards (MIL-STD, ANSI or company specifications).

The schematic diagram is a dimensionless drawing, that is, there are no dimensions for the drafter to follow except those that are determined at the onset or according to the following procedures:

Grid-Paper Estimating. One method is to use grid paper to solve both the layout and final diagram problem. The grid paper is available in both roll and standard sheet sizes. The standard grid spacings are 8x8 or 10x10 squares to the inch, with each one inch division printed in heavier lines. The grid paper is printed in a fade-out blue ink which will drop out when a blueprint is made of the original.

The procedure is to assign arbitrary spaces to the symbols on the basis of grid spacing. An example: (Figure 1-22) a symbol that is eight grids long would be rounded off at 10 grids, thus a common basis is established for all symbols of that variety. Therefore, some dimensions and spacing can be used in estimating the total area needed for the finished diagram. Figure 1-23 shows components laid out on grid.

The use of grid paper has many time saving advantages in the trial layouts and final diagram. Once the drafter has made one or more trial layouts on grid paper, it becomes easy to segregate major sections of the circuit and thus arrive at the best possible space saving arrangement. These rough trial layouts can be made easily on grid paper and thus eliminates the need to make major changes on the final diagram.

When using grid paper, it is recommended that the drafter use a template that is made for that particular grid spacing. It is also recommended that the 8x8 grid be used when drawing schematic diagrams.

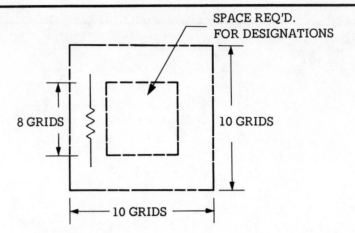

FIGURE 1-22

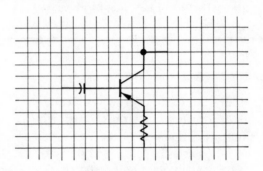

FIGURE 1-23

Estimating Graphic Symbols by Number. Another method in determining the size of the format, is to count the total number of graphic symbols required for the schematic diagram. Table 1-2 illustrates the relationship between total number of symbols needed and the drawing area. The table is used as an estimating guide only. The table includes considerations for reference designations, notes, component values and other data needed on the schematic diagram.

The table does not allow for the subdividing of large schematics into small units which would require more space, or for a schematic diagram that requires a large number of interconnecting lines between circuits, or one that has a large number of signals that run from side to side. Hence, this also reduces the possible number of symbols that can be located on a particular drawing size.

TABLE 1.2   Drawing Size Compared with Number of Graphic Symbols

	Drawing Size	Drawing Area (sq. in.)	Number of Symbols Allowing 2½ sq. in. per Symbol
B	11 × 17	187	55
C	17 × 22	374	120
D	22 × 34	748	250
E	34 × 44	1496	525

SCHEMATIC DIAGRAM DRAWING CONSIDERATIONS

The layout of the schematic diagram can be completed more efficiently if a predetermined procedure is followed. This technique applies whether the circuit is large and complex or small and simple. By following the procedures described below, it will not only make the drawing easier, but will also result in a better end product. These guidelines should be followed when a clear and well arranged schematic is required—in other words, always:

1. Check with the engineer to see that all the circuit data and required documentation is clear and understood.

2. Check to see if the circuit is subdivided into individual circuits, and if it is, what identifications are needed. Also ask the engineer if any revisions have to be incorporated, or if space is required for possible increases in circuitry, and if any extra space should be allotted.

3. Make a freehand sketch of each schematic circuit on grid paper. Allow enough space around the symbol for reference data and also enough space between the adjacent symbols. This step cannot be overly emphasized. Make several layouts in order to achieve the best balance and flow.

4. Make freehand sketches of any auxilliary circuits. Examples: power supply or complex schematic symbols.

5. Make a list of all general notes and tables needed, and establish approximate area required.

6. Make a combined sketch of all circuits, notes and tables needed for the final diagram. Use grid paper to help in spacing. The combined sketch should be sketched within the drawing format size. Caution should be taken to avoid the revision block, title block and other preassigned areas on the drawing.

7. Check to see that all the symbols are correctly drawn, that correct signal flow is observed and a pleasing balance is obtained.

8. Check to see that any similar circuits are drawn uniformly.

9. Check to see that the ground symbols are located near the bottom and the voltages are located near the top of the diagram.

10. When more than one sheet is used, make sure that all signals are clearly identified on each sheet, and if the signals run to an adjacent sheet, the circuit connects should match if possible (Figure 1-24).

11. Test points are to be clearly marked.

12. Check to see that all inputs and outputs are identified.

13. Using the revised rough sketch, proceed in making the final schematic diagram on 8x8 grid paper. If the company format isn't gridded, slide a piece of grid paper under this format. This will expedite the drawing process.

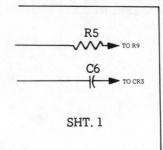
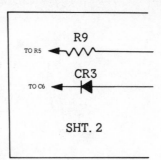

SHT. 1   SHT. 2

FIGURE 1-24

The arrangement of the symbols and the interconnecting signal paths on the schematic diagram should present a uniform and consistent appearance. The symbols should be functionally arranged to allow for the most direct routes, avoiding unnecessary crossovers and corners. Figure 1-25 illustrates some ways of simplifying the diagram. An important point to remember is that caution should be exercised when rerouting signals on the diagram for esthetics; signal-connection accuracy is the primary objective. Without proper electrical connection, the diagram is of no value to the person trying to interpret it.

Graphic symbols should be drawn either vertically or horizontally. A pleasing balance can be achieved when the symbols are aligned with each other. Figure 1-26 illustrates some typical arrangements.

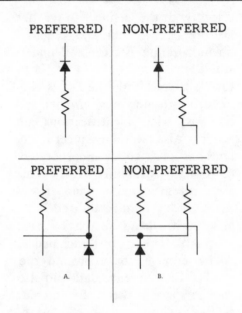

FIGURE 1-25

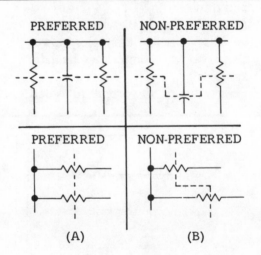

FIGURE 1-26

SYMBOL AND CONDUCTOR SPACING

Conductor paths on the schematic should be a minimum of 2 grids or ¼ inch apart. This allows enough room between lines if the diagram is to be reduced for microfilm storage. If a number of signal paths run parallel to each other, it is recommended that the signal paths be grouped, preferably three to a group, with approximately double spacing between groups. An example would be if the normal signal spacing is .25 between lines, then the spacing between groups would be .50 (Figure 1-30). This makes the diagram easier to read and less confusing to the eye.

Conductor paths that run parallel (with graphic symbols included) should be spaced with a minimum of 1.25 " between symbols and traces. This allows ample room for the identifications. Figure 1-27 depicts a typical condition.

When placing a graphic symbol within a conductor path line, allow a minimum of .375 " from a tie point or a crossover, Figure 1-28. If two or more graphic symbols are placed on the same conductor line, allow a minimum of .75 " between the adjacent symbols, Figure 1-29. This space will allow the necessary room for identification. The spacing dimensions given for schematic diagrams are recommended minimums, and may vary with individual company requirements.

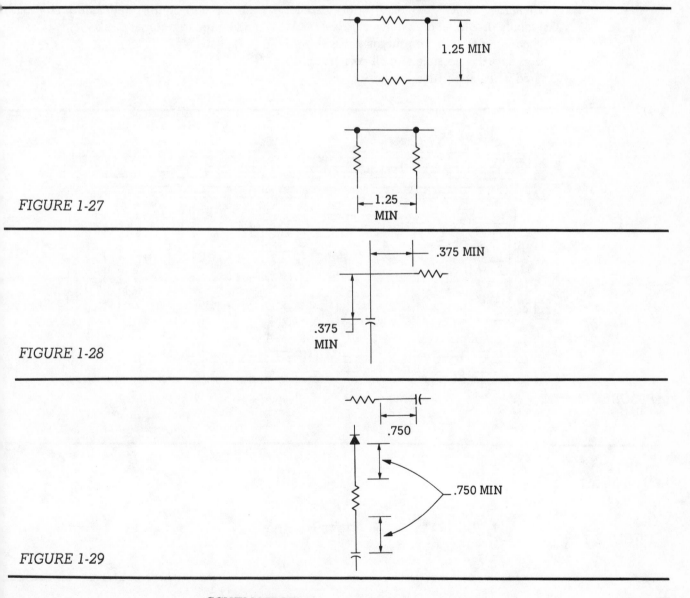

FIGURE 1-27

FIGURE 1-28

FIGURE 1-29

SIGNAL CONNECTIONS AND CROSSOVERS

The junctions of signal connections can be shown using either of two methods. Figure 1-31A illustrates the technique of using a dot to show an electrical connection, and Figure 1-31B illustrates the no-dot method of an electrical connection. Both methods are acceptable; however, only one technique should be used on a schematic diagram. When using the dot method, it is recommended that no four way ties be used. (See Figure 1-31C). This prevents the possibility of error caused by accidental omission of the dot which could give the appearance of a crossover instead of a connection (Figure 1-31D).

CONDUCTOR PATHS SIMPLIFIED

An engineer's freehand sketch of the schematic will often show parallel symbols; example: diodes, resistors and capacitors. Figure 1-32B illustrates this arrangement. To eliminate some corners and make the schematic easier to read, the parallel symbols should be drawn per Figure 1-32A. Whenever possible, conductor paths should be as short as possible and all corners eliminated when applicable.

The engineer's freehand sketch of components that connect to ground is usually indicated by the illustrations in Figure 1-33B. The preferred method is illustrated in Figure 1-33A. This technique provides shorter and more direct conductor paths and eliminates corners.

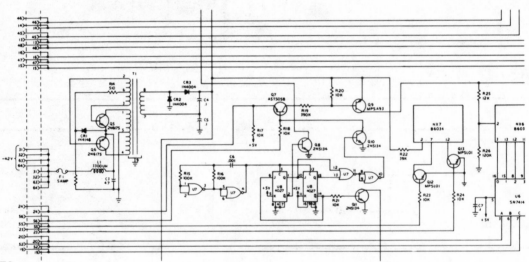

FIGURE 1-30

FIGURE 1-31

(A)   (B)   (C)   (D)

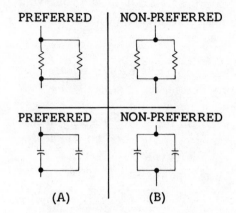

PREFERRED | NON-PREFERRED

PREFERRED | NON-PREFERRED

(A) | (B)

FIGURE 1-32

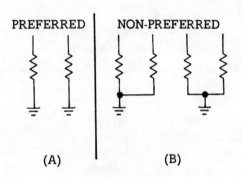

PREFERRED | NON-PREFERRED

(A) | (B)

FIGURE 1-33

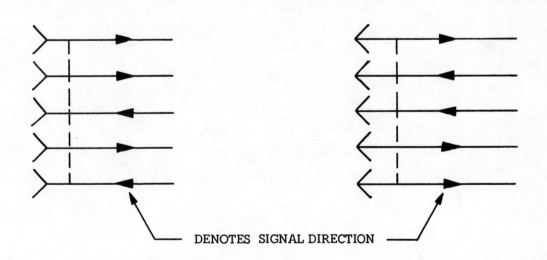

FIGURE 1-34

DENOTES SIGNAL DIRECTION

SCHEMATIC CHECK LIST

After the drafter has completed the schematic diagram (which meets either the MIL, ANSI or company standards) one final check must be made of the following items to insure accuracy and completeness in the schematic diagram:

1. Before the schematic is released for production, the engineer will review the schematic for flow and arrangement of the logic or components.
2. Primary signal lines on all analog schematics will be drawn heavier.
3. All test points should have a name.
4. The schematic should be drawn with "–" voltages down and "+" voltages up.
5. If possible, show the **PC** board components and/or controls with their related functions.
6. Do not draw power and ground lines on the schematic unless it is very easy to see. Use the symbols and codes.
7. All inputs and outputs on the connector should show the signal direction: see Figure 1-34 for an example of this technique.
8. Switches should be marked N.C. (normally closed) or N.O. (normally open).
9. IC's use reference designation—"U".
10. "OP-AMPS" eliminate all dash lines around the symbol.
11. IC's with gates that have internal connections should have a box drawn around the logic.
12. Decoupling logic capacitors should be placed in the notes only.
13. Coding inputs and outputs from gate to gate use a number "(X)" to indicate the number of places it terminates. A note will be required to define (X).
14. IC's list all input and output coding.
15. When possible, show internal logic on all blocks drawn on the schematic.
16. Dash lines shown on the schematic must be defined in the notes.
17. Resistors and capacitors values should be complete. Example: 200K, 1/4W, 5% (or in the notes).
18. Zener diodes should have the voltage rating and tolerance.
19. Spare gates should be shown.
20. Transformers—the nominal operating voltage on all output terminals should be listed.

Chapter

2

BOARD LAYOUT GROUND RULES

The layout is undimensioned except for the board size and its relationship with mating structures and assemblies such as loading holes or special mounting fixtures. Tolerances should be included to facilitate further design and drafting.

Board layout and artwork should be made to the same scale whenever practical. After completion of the layout, an art-master (tape-up) is generated. The art-master is made to an enlarged scale (generally 2:1 or 4:1 depending on a particular company's standards). It is then photo-reduced by a photographer to provide a 1:1 scale negative and positive (master pattern) which becomes a highly accurate tool for the manufacturing of the P. C. Board.

Other documentation prepared from the layout and the reduced positive include fabrication, assembly, parts list and silk-screen drawings, plus whatever special manufacturing sketches or aids that might be required. These are shown in Table 2-1.

Part numbers or other suitable identification such as component reference designations, test point designations, connector pin numbers, etc. taken from the schematic should be provided to identify all items and components appearing on the layout. The initial layout may start by arranging a rough sketch of the circuitry to determine the most practical placement of components and wiring (traces). The next step is to adopt the sketch to a formal layout using a grid and a component template which will establish very accurate conductor paths, component body outlines, component pad sizes, board outline, connector size and all other features. The layout should be drawn as viewed from the component side of the Printed Circuit Board. The grid pattern should be .050 or .100 inch spacing as shown in Figure 2-1.

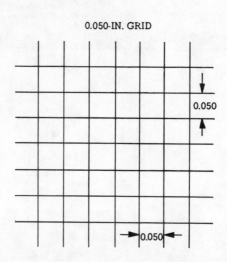

0.050-IN. GRID

0.050

0.050

FIGURE 2-1 Grid Layout Pattern for PCB's

1.0 Printed Circuit Documentation

Whether a PC board is to be a one-of-a-kind prototype, or a high-volume production article, it should have some informative type of documentation describing the means to the end. A printed-wiring drawing package should include the drawings shown below. Exactly *how much documentation,* and how it should be prepared will vary with budget and schedule as well as application.

Adequate documentation conveys to the user the basic electromechanical design concept, the type and quantity of parts and materials required, special manufacturing instructions, and up-to-date revisions.

Too little documentation results in misinterpretation, information gaps, and loss of uniform configuration. Manufacturing becomes dependent on individuals rather than documentation, often resulting in expensive rework and valuable lost time.

Too much documentation can result in increased drafting costs, and decreased manufacturing productivity due to

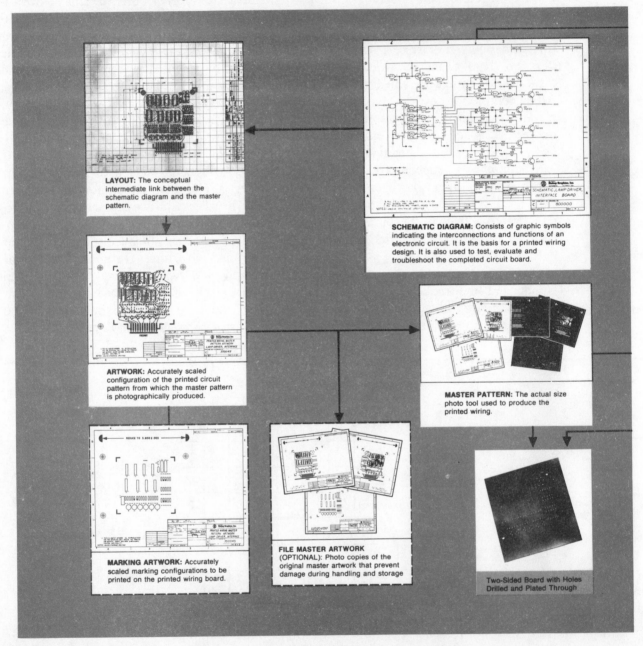

LAYOUT: The conceptual intermediate link between the schematic diagram and the master pattern.

SCHEMATIC DIAGRAM: Consists of graphic symbols indicating the interconnections and functions of an electronic circuit. It is the basis for a printed wiring design. It is also used to test, evaluate and troubleshoot the completed circuit board.

ARTWORK: Accurately scaled configuration of the printed circuit pattern from which the master pattern is photographically produced.

MASTER PATTERN: The actual size photo tool used to produce the printed wiring.

MARKING ARTWORK: Accurately scaled marking configurations to be printed on the printed wiring board.

FILE MASTER ARTWORK (OPTIONAL): Photo copies of the original master artwork that prevent damage during handling and storage

Two-Sided Board with Holes Drilled and Plated Through

TABLE 2-1

time-consuming interpretation of overly complicated and confusing drawings.

Printed wiring documentation may be divided into three classifications:

1. *Minimum—used for prototype and small quantity runs.*
2. *Formal—used for a standard product line and/or production quantities. Similar to Category E, Form 2 per MIL-D-1000 without source or specification support documentation.*
3. *Military—complies with government contracts specifying procurement drawings for the manufacture of identical items by other than the original manufacturer. (Category E, Form 1, per MIL-D-1000.)*

Standard format and reprographic drafting techniques are used extensively to reduce costs and improve drawing readability.

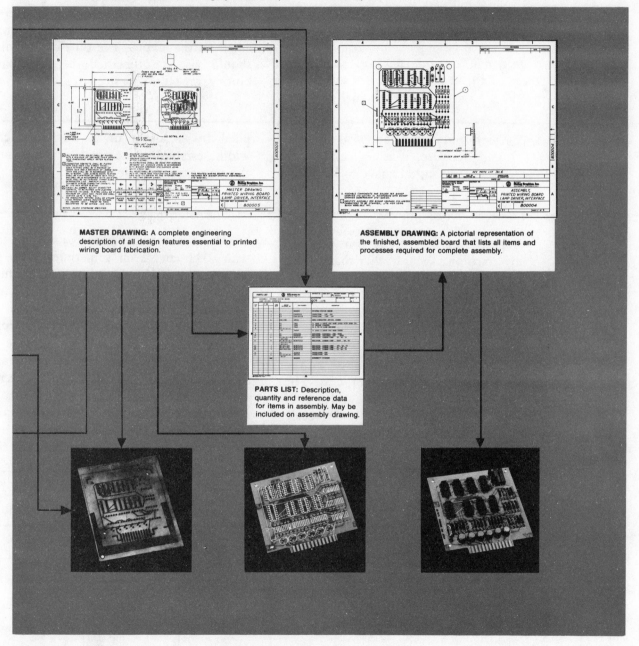

MASTER DRAWING: A complete engineering description of all design features essential to printed wiring board fabrication.

ASSEMBLY DRAWING: A pictorial representation of the finished, assembled board that lists all items and processes required for complete assembly.

PARTS LIST: Description, quantity and reference data for items in assembly. May be included on assembly drawing.

The board layout should be made to an enlarged scale. Either 2/1 or 4/1. All component pads should be located on grid intersections. The pad spacing for components should be determined by the maximum component body size. See standard formula in Chapter 4. Components should be mounted on only one side of the board. Components should be placed or located so that any component may be removed without removing any other component unless the assembly is not intended to be repaired.

Only one component lead should be placed in a lead hole except for some commercial bus bars that are designed to fit in the same mounting as the integrated circuit lead. When a board design requires that jumpers are to be used, (a jumper is a insulated wire used to substitute a trace on the board) they should be placed on the component side of the board and constructed of solid insulated wire designed to be as straight and as short as possible. Whenever possible avoid crossing over components. Components requiring special hold-downs because of shock, vibration or weight considerations should be so indicated on the layout. The type of fastening should also be specified; i.e., clamping, bolting, rat-tails, adhesive, etc.

CONDUCTOR WIDTH

The preferred board layout is done at 2/1 scale; therefore, it is logical that all traces, pads, air gap, etc. have to be multiplied by a factor of 2. For conductor width, establish a nominal size that suits the majority of applications.

SIGNAL TRACES

A signal trace is generally considered to be any trace on a board other than power and ground. These traces are generally low voltage and low current and, therefore they do not require a massive trace or air gap for proper function.

Traces on the PCB are produced by special adhesive tape that is applied to the artmaster during the design process. The most commonly accepted tape width used in industry today is .040 in. at 2:1 scale with .040 in. air gap between adjacent traces or pads. In cases of extremely dense boards, it is not uncommon to see tape-ups with .031- and even sometimes .026- in. wide traces; but it should be kept in mind that as the trace and air gap decrease, the cost and unreliability of the board increases. The .040-in. trace and air gap combination mentioned above provide a .020-in. conductor on the actual board. This combination provides ample material for conduction of electricity and also allows sufficient tolerance for the board manufacturing from the initial tape-up through the photography, etching, plating

TAPE WIDTH IN INCHES

.015	.050	.125	.187	.312	.800
.020	.062	.140	.200	.375	1.000
.026	.070	.150	.218	.400	2.000
.031	.080	.156	.250	.500	
.040	.093	.160	.280	.750	
.046	.100		.300		

TABLE 2-2

and wave soldering processes. Each of these steps should be allowed an acceptable amount of tolerance so that cumulative errors do not affect the reliability of the completed board.

In cases where the trace width (tape) must go down to .026- or .031-in. widths, it becomes very critical to try to maintain at least .040-in. air gap instead of an air gap the same width as the tape as illustrated in Fig. 2-2, the air gap should be adequate to avoid solder bridging between traces during the wave soldering operation.

POWER/ GROUND TRACES

The primary consideration in a power or ground trace is to provide a direct connection from the device to the power supply. Because this is not always possible, the next best step is to increase the width of the conductor trace so as to provide enough mass to accomplish essentially the same thing.

As shown in Fig. 2-3 a nominal trace width for power and ground can be established at .100-in. (2:1 scale). As in the case of the signal trace, this can be reduced for some requirements, but be sure the electrical engineer is aware of this condition. A .100-in. trace is also easy to plan as the design progresses due to the fact it is an even grid width and the air gap could be the same.

Color coding can be used in three differ-

ent methods to provide helpful aids to the P.C. designer; i.e., conductor layer coding, voltage/ground pads, and input/output pads. Although each color coding method serves a separate distinct purpose, they are all used during the course of a typical P.C. layout.

A standard colored pencil such as those available from office supply or drafting supply outlets can be used for each color. Pick a readily available color to represent a particular voltage, ground or trace. Become familiar with what it represents and do not change. This avoids the confusion of what each color means on each board. After using these approaches for awhile, you will automatically relate the color code to its function without referring to a color key.

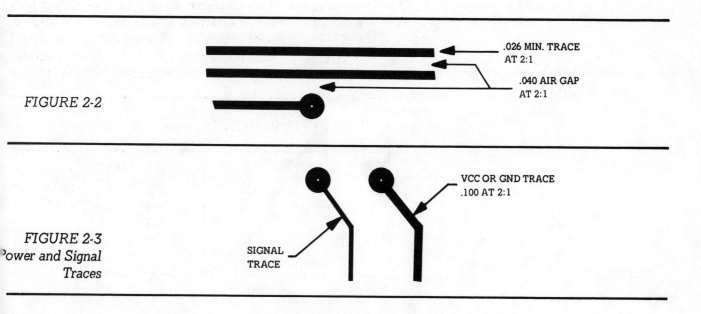

FIGURE 2-2

.026 MIN. TRACE AT 2:1

.040 AIR GAP AT 2:1

VCC OR GND TRACE .100 AT 2:1

SIGNAL TRACE

FIGURE 2-3
Power and Signal Traces

CONDUCTOR LAYER CODING

This coding system indicates which side of the PCB a particular trace is located. Therefore, on a typical two-sided board, two different colors would be used; i.e., red traces indicate component side traces, blue traces indicate non-component side traces. The component body outlines would be drawn with yet another color such as black or some color that you would not plan to use for any of the other color coding methods. If a board is intended to be single-sided (all conductors on one side of the board only), then all traces should be one color only. See Figure 7-9

VOLTAGES/GROUND CODING

In this method, a different color is chosen to represent each voltage and ground that is used; i.e., red might be +5V,

orange +12V, green –12V, and blue ground. This is accomplished by color coding the component lead pad that is to be connected to the particular voltage or ground. (Pad only)

During the design stage, the colors stand out as flags to indicate where the nearest voltage or ground point is located for the most direct hook-up. This keeps the traces very direct and in most analog boards, allows the designer to wait until the layout is near completion to hook all the respective traces together.

During the taping operation, the color coded pad on the layout can be seen through the hole in the pad on the artwork. As shown in Figure 2-5 a general rule would tell the PCB taper that all color coded pads were to be .100-in. tape minimum regardless of the color; all other traces could be .050-in. wide or whatever size you choose.

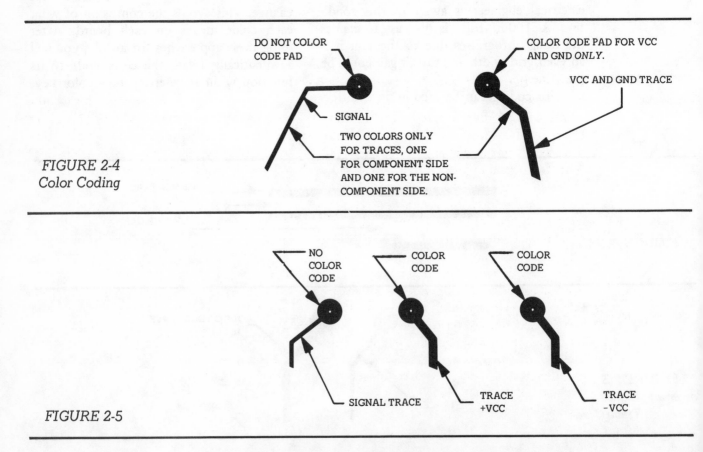

FIGURE 2-4
Color Coding

FIGURE 2-5

BOARD LAYOUT GROUND RULES

OUTPUT CODING

Terminals (pads) that are to be connected to a connector(s) may also use a color coding method. As the circuit is being laid out, the particular pad that is to go out to the connector can be color coded and left to be routed after the rest of the layout is completed. This allows the designer to complete the connections between local related components and then route the output connections (traces to the connector) in whatever path is most available and less crowded. The colors used for this purpose would be different from those used for voltage/ground coding and each different color would represent a particular connector; i.e., brown - J1, violet - J2.

TEMPLATES

A standard component template is one of the most necessary pieces of equipment the designer will use in P.C. design. It not only saves valuable time but also assures that standard component lead spacing and body sizes are considered. Although several models and styles are available, pick one that will be compact, yet complete for the requirements of your particular type of design. See FIG. 2-6

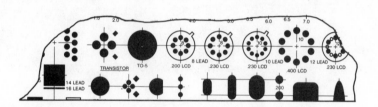

FIGURE 2-6

FEED THROUGHS

A feed-through consists of a pad on both sides of the board with a plated-through hole connecting them. They are used to connect traces on one side of the board to traces on the opposite side when a component pad is not available to accomplish this purpose. Feed-through holes are constructed like any other plated through holes; but because no terminals or component leads are mounted in them, they are generally smaller than any other pad and plated-through hole in order not to consume excess space on the board that is needed for trace routing. Feed-throughs are especially necessary in digital layouts where vertical and horizontal traces must be kept on opposite sides as much as possible. This method requires that a trace be routed in one direction as far as possible, then a feed-through pad located and the connection is continued on the opposite side of the board.

Feed-through sizes vary depending on the density of the board; but on tape-ups (where practical), feed through pads should be .125-in. This results in a .062-in. pad on the actual board which could have a .025- to .031-in. diameter hole drilled through it.

Feed-through pads smaller than .125-in. are used in many cases; i.e., .090-, .100-in., but a hole smaller than .025-in. diameter becomes impractical for the board manufacturer to drill in quantities. Therefore, the smaller the diameter of the pad, the narrower the annular ring (see page 62) will become on the finished plated-through pad.

In general, the pad size requirements and the hole size requirements for a feed-through hole should be held to the same standards and specifications of any other plated-through hole.

Chapter

3

DISCRETE COMPONENT LAYOUT

DISCRETE COMPONENTS LAYOUT (ANALOG BOARDS)

The following steps should be taken prior to laying out a PCB. These will simplify the design process.

1. All symbols or reference designations used by the engineer should be understood.

2. The specification sheet must be used to identify the case size or the body width and length (maximum size).

3. The case size should be used to determine the lead spacing to mount the component on the P.C. board (See following for minimum dimensions over maximum case size for lead spacing.) See Chapter 4, page 57

4. Polarity must be determined for all polarized components such as capacitors, diodes (must identify the cathode end) and some transistors (must identify the emitter, base and collector).

5. The lead diameter must be determined in order to select a pad size on our layout. See Chapter 4, page 62

6. Determine pad orientation looking from top view of components such as the transistor's emitter, base and collector, potentiometer or variable resistor Pins 1, 2 and 3, variable capacitors, relays, transformers, coils, etc.

7. Convert or establish component sizes at a 2/1 or 4/1 scale, which will be the same size of the layout. At this point, try to find most component sizes on the P.C. design template; for example, ¼W resistors, ½W resistors, diodes, ceramic disc capacitors, T0-5 and T0-18 case transistors, and many more depending on which template you use.

8. Before using the schematic to electrically group and connect components, look at the space requirements. This will conserve space relative to the different size components on a typical P.C. board. Figure 3-1 shows how important arrangement versus size can be.

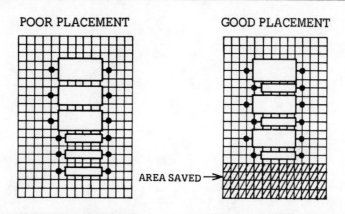

POOR PLACEMENT   GOOD PLACEMENT

AREA SAVED →

FIGURE 3-1
Parts Placement

9. The schematic in Figure 3-2 will be used to illustrate the grouping of discrete components.

a. Assuming steps 1 through 8 have been considered for component sizes, check to see that all components have the proper reference designators.

b. The initial layout may start by arranging a rough arrangement of the components to determine the most practical placement. The component could be drawn in a rough form; use the template at this point, see example of Figure 3-3 using the schematic of Figure 3-2.

10. The "Correct Method" of Figure 3-3 is just one way to group the components in Figure 3-2: there are several ways that would be considered correct. One point it does show is that a component must be selected to work around or build from such as transistor Q1. In this case, the transistor is the key component because it connects to more components than any other one component on the schematic. Another point to consider is the shape of the grouped components. Try to group components in a square or rectangular shape. Refer to Figure 3-4 "Incorrect Method"; this shape is difficult to work with as far as adding circuits to the board at a later date or

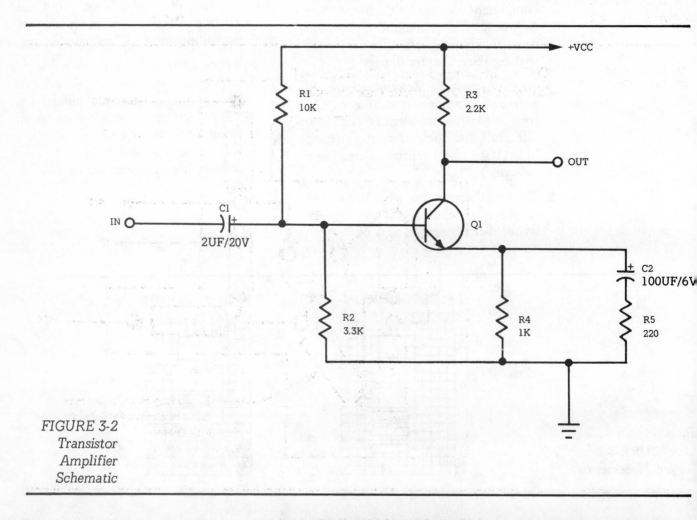

FIGURE 3-2
Transistor
Amplifier
Schematic

DISCRETE COMPONENT LAYOUT

working additional components around this shape.

11. The layout of Figure 3-5 is electrically the same as Figure 3-4. With the same electrical connections, the components can be arranged differently.

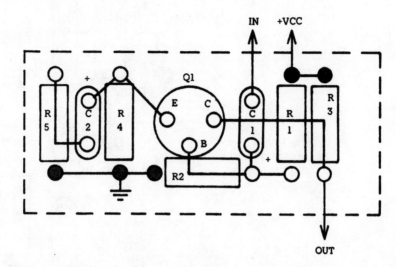

● PADS = +VCC
● PADS = GND

FIGURE 3-3
Correct Layout
for Schematic
Shown in Fig. 3-2

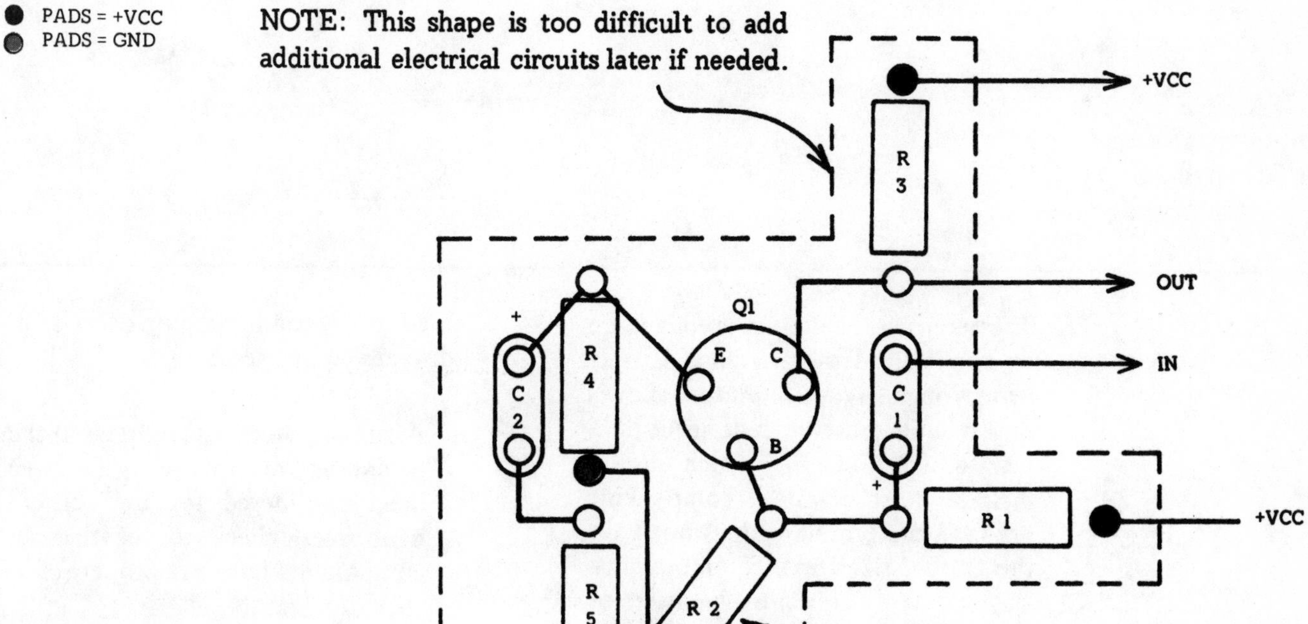

● PADS = +VCC
● PADS = GND

NOTE: This shape is too difficult to add additional electrical circuits later if needed.

This is very bad. Do not turn components less than 90° to each other.

FIGURE 3-4
Incorrect Method
for Schematic of
Figure 3-2

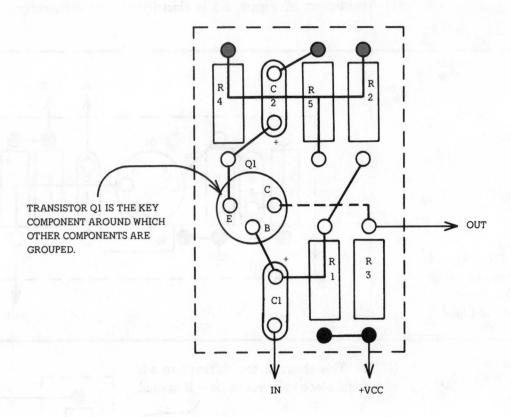

● PADS = +VCC
● PADS = GND

TRANSISTOR Q1 IS THE KEY
COMPONENT AROUND WHICH
OTHER COMPONENTS ARE
GROUPED.

FIGURE 3-5
Another Layout
for Schematic of
Figure 3-2

12. The choice of the final layout is up to the PCB designer. Once a free-hand rough layout (not to scale) is selected, it must be redone using a P.C. design template which should include most common components used. If the template does not have the correct size for the components, a circle template must be used to construct the component to "scale." Be sure that you select or construct it to the maximum component size. For further information about formal layout, see the P.C. board layout section.

13. The schematic of Figure 3-6 will be used as a second grouping exercise of discrete components.

a. Assuming Steps 1 through 8 in the beginning of this chapter have been considered for the component sizes, check to see that all components have the proper reference designators.

b. Again as in Step 9b, the initial layout should start by arranging a rough layout of the components to quickly determine the most practical and effective placement electrically and mechanically.

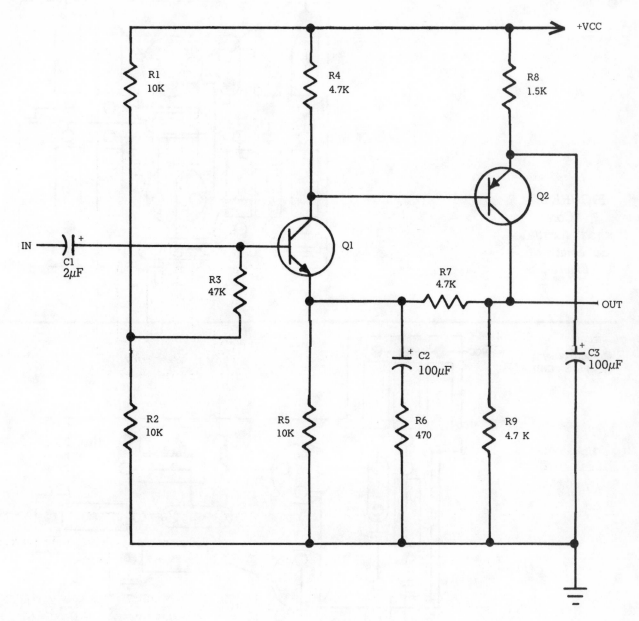

+VCC

R1
10K

R4
4.7K

R8
1.5K

Q2

IN

C1
2μF

R3
47K

Q1

R7
4.7K

OUT

C2
100μF

C3
100μF

R2
10K

R5
10K

R6
470

R9
4.7 K

FIGURE 3-6 Two
Transistor
Amplifier
Schematic

14. Figure 3-7 is a rough layout of the Correct Method for the Schematic Figure 3-6.

15. The layout of Figure 3-8 shows the

wrong way to spread out components. By *not* grouping in a square or rectangular shape, you *wasted* area.

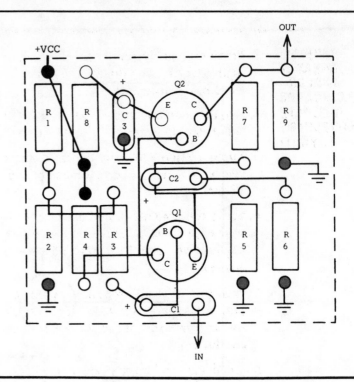

FIGURE 3-7
"Correct
Method" for the
Schematic of
Figure 3-6

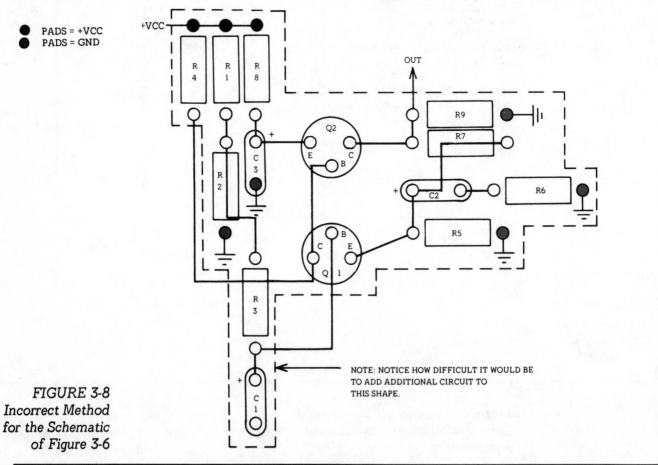

NOTE: NOTICE HOW DIFFICULT IT WOULD BE
TO ADD ADDITIONAL CIRCUIT TO
THIS SHAPE.

FIGURE 3-8
Incorrect Method
for the Schematic
of Figure 3-6

PROBLEM AREAS OF ANALOG— P.C. BOARDS (DISCRETE COMPONENT LAYOUT)

1. Predetermine location of all components which are dictated by special structural considerations.

2. Arrange components so as to achieve the optimum shape for the P.C. board.

3. Some components should have short traces or interconnects (engineer will advise).

4. Some components should not be placed near each other to prevent magnetic or capacitive coupling (engineer will advise).

5. High potential differences between certain conductors will make it necessary to increase the space between such conductors.

6. Metal case (without insulator) components should not touch each other. Case could be electrically hot.

7. Separate grounds could be used on the same schematic. (Engineer will advise.)

8. Placement of variable components on the P.C. board should be considered for ease of access.

9. Size of voltage and ground traces.

10. Signal traces should be considered in the layout before voltage or ground traces. (Typically the voltage or ground traces can be as long as you wish).

11. Ground planes are very common on analog P.C. boards and should be considered in the initial layout. Some components use ground patterns for electrostatic shielding.

12. Some components require heat sinks (very common with transistors). (Engineer will advise).

13. Locate heat-sensitive and heat-radiating components as far apart as practical. But try not to locate the heat-radiating components together.

14. Components weighing ½ oz or more should not be supported by the leads solder joint alone, but should be provided with additional means of support such as brackets or clamps.

Chapter

4

MOUNTING COMPONENTS

Component mounting is probably the most important step in the printed circuit board design phase. The importance is simply that a specific number of components must be mounted onto a P.C. board with specific dimensions. This is complicated by some pretty rigorous spacing standards which make the job of mounting a seemingly difficult one. However, by following some basic rules, the job can be made fairly uncomplicated.

One of the most important steps the designer must perform is to determine how an actual component will be mounted into a finished P.C. board. Try to imagine an assembler trying to stuff resistors with .025-in. diameter leads into .020-in. pad holes. The consequences are dreary. Follow the steps below and this should never happen. Figure 4-1A and B show the details of component mounting.

The three steps in component mounting are to determine pad spacing, P.C. board hole (drill) diameter, and pad diameter.

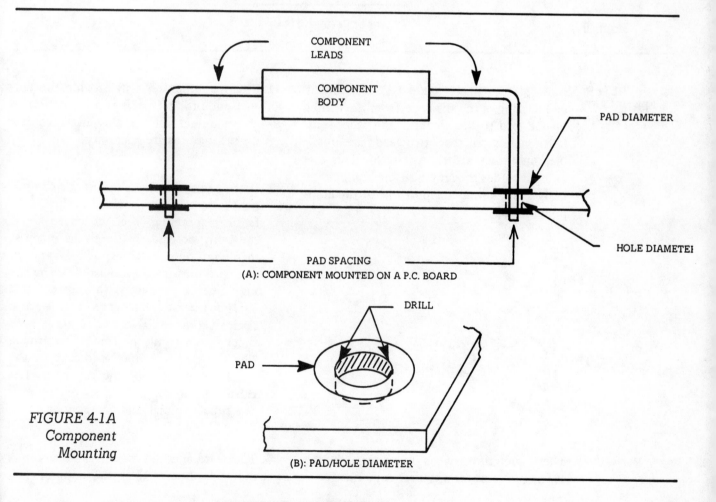

FIGURE 4-1A
Component
Mounting

(A): COMPONENT MOUNTED ON A P.C. BOARD

(B): PAD/HOLE DIAMETER

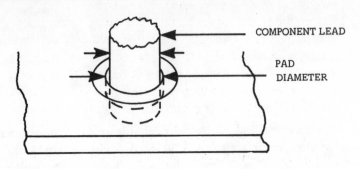

COMPONENT LEAD

PAD DIAMETER

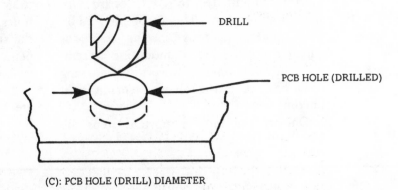

DRILL

PCB HOLE (DRILLED)

FIGURE 4-1B
Component
Mounting

(C): PCB HOLE (DRILL) DIAMETER

DETER-MINING PAD SPACING

As shown in Figure 4-1A, component pad spacing is the distance from the center of one lead to the center of the other lead. (Although more than two leads are possible, only two are shown here). This distance includes the length of the component body plus the unbent portions of the component leads.

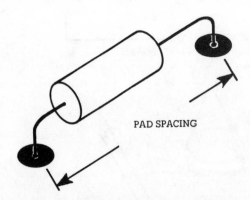

PAD SPACING

There are three ways to determine the spacing of pads for a typical component:

1. The manufacturer will provide the correct mounting (spacing) information in the manufacturer's catalog, in which case no calculations are necessary.

2. If the mounting (spacing) of the component's leads is not provided in the manufacturer's catalog, then check to see if the body or case size is given. In most manufacturer's catalogs, the maximum case size dimensions are listed. Using the maximum dimensions, the correct pad spacing can be determined. Fig. 4-4 is a typical commercial manufacturer's spec sheet which would be found in a catalog. The size can be found by looking up the voltage and value of the capacitor. The capacitor shown is just an example of one type of radial lead component.

3. The third method is not as accurate as methods 1 and 2. If the component's

MOUNTING COMPONENTS

mounting or hole spacing cannot be found from the manufacturer's catalog or the case size (for axial or radial components), the actual part must be scaled. Try to pick up a sample from the manufacturer or distributor. Once the part is obtained, then use the same formula as in method 2 and determine the pad spacing by scaling the part to get the maximum body length required on an axial lead component or just the distance between the leads for a radial lead component.

HOLE DIAMETER & PAD SIZE

Hole diameter and pad size depend on each other. Using the formula listed below, the hole size must be determined before a pad size is selected. To calculate the hole diameter, the maximum lead diameter of the component to be inserted must be known.

Lead Diameter: Finding the lead diameter of a component can be done in two ways: First by looking in a manufacturer's catalog and finding the maximum diameter as listed; second by obtaining the actual part and scaling the lead diameter. The first choice is by far more accurate.

Hole Diameter Formula: .006-in. minimum to .020-in. maximum over the maximum lead diameter = the hole size as shown in Figure 4-5A.

Pad Spacing Formula: In order to use this formula, the case size of the component (axial lead type) must be known. Fig. 4-2.

Take the maximum body (case) size length "X" + .060-in. minimum on each side from body to centerline of each lead = pad spacing (This will give the minimum spacing between pads). An important point to keep in mind is that this formula gives the *minimum* spacing. Because most components are laid out on even tenth or .100-in. spacing at 1/1 scale, this is called or referred to as laying out on "grid."

Example: If calculations for the spacing of a component worked out to be ".460 thousands" at 1/1 scale, round this off to ".500" at 1/1 or "1.000" at 2/1 scale.

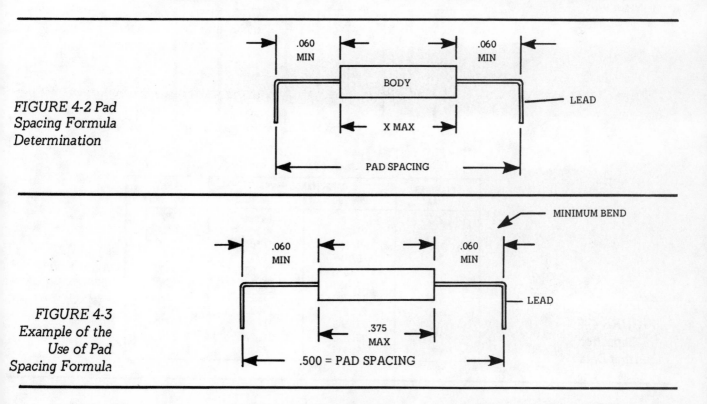

FIGURE 4-2 Pad Spacing Formula Determination

FIGURE 4-3 Example of the Use of Pad Spacing Formula

Working Exercise-See Figure 4-3

X = .375-in. maximum + .060-in. x 2 = pad spacing

.375	body length
.060	minimum bend
+.060	minimum bend
.495	minimum pad spacing

The .495 dimension would work, but to stay on grid, we would round this off to .500 at 1/1 or 1.000 at 2/1 scale.

NOTE: When rounding off axial lead type components, try to round off in these increments at 1/1 scale: .200, .250, .300, .400, .500, .600, .700, .800, .900, and 1.000.

Radial Lead Components: Radial lead components have no standard formula such as the axial lead components. Instead, you must rely on the manufacturer to supply you with the proper hole or pad spacing. See example in Figure 4-4.

Pad Size Formula: Minimum .020-in. annular ring over the maximum hole size (.020 annular ring is for reference only)

NOTE: Applying the formula of .020-in. minimum annual ring over the maximum hole size will give the minimum size pad that could by used. This minimum size pad diameter may not be a standard size pad that can be bought. If this is so, simply round off to the next higher standard pad size.

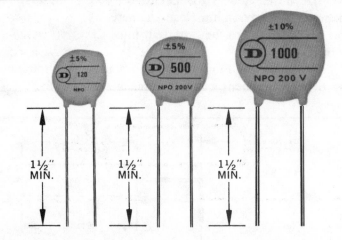

FIGURE 4-4
Capacitor
Mounting Data

DIAMETER MAX.	CAPACITANCE RANGE		LEAD SPACING*
	200 VDC	500 VDC	
.225″	1-22 pf	1-18 pf	.187″
.285″	23-43	19-30	.250″
.345″	44-68	31-50	.250″
.385″	69-91	51-75	.250″
.440″	92-130	76-100	.250″
.520″	131-150	101-135	.375″
.575″	151-200	136-150	.375″
.675″	201-250	151-210	.375″
.735″	251-300	211-240	.375″
.770″	301-350	241-270	.375″
.850″	351-500	271-360	.375″

*ROUND OFF
TO THE NEXT
HIGHER GRID
(.100 GRID)

TYPICAL AXIAL LEAD COMPONENT

Step 1—The manufacturer's spec sheet Figure 4-5 lists the lead diameter as .019-in. with a tolerance of ±.002″. Now add the tolerance ±.002 to the lead diameter .019:

```
 .019  Lead diameter
+.002  Tolerance
 .021  Maximum lead diameter
```

Step 2—Once the maximum lead diameter is established, add the minimum clearance which is .006-in. to the maximum lead diameter.

```
 .021  Maximum lead diameter
+.006  Minimum clearance around lead
 .027  Minimum hole size
```

Step 3—Once the minimum hole diameter is established, find the maximum hole diameter. The maximum clearance over the lead is .020-in.

```
 .021  Maximum lead diameter
+.020  Maximum clearance over the lead
 .041  Maximum hole diameter
```

Step 4—Two hole sizes (minimum and maximum) have been established. Use this to find which size hole would best fit as many other components as possible on the P.C. board layout (See example below).

Example: Compare the minimum and maximum hole diameters of component "A" to component "B" to see if a hole size can be selected that will fit both components Figure 4-5. This can be done by examining the range of hole sizes for components "A" and "B."

Component "A" holes sizes range from a minimum of .027- through .041-in.

Component "B" hole sizes range from a minimum of .034- through .048-in.

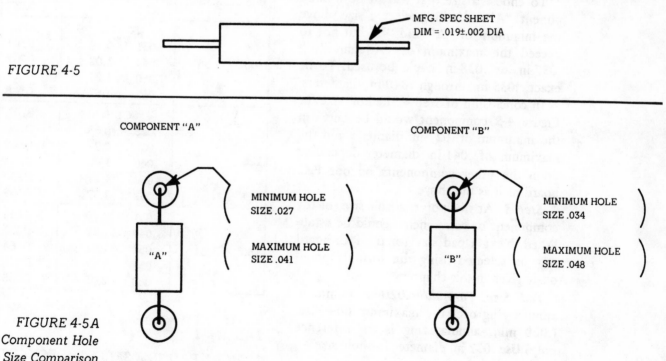

FIGURE 4-5

MFG. SPEC SHEET
DIM = .019±.002 DIA

COMPONENT "A"

MINIMUM HOLE SIZE .027

MAXIMUM HOLE SIZE .041

"A"

COMPONENT "B"

MINIMUM HOLE SIZE .034

MAXIMUM HOLE SIZE .048

"B"

FIGURE 4-5A
Component Hole
Size Comparison

Min. .027
Comp. .028
"A" .029
.030
.031
.032
.033

Min. .034
Comp. .035
"B" .036
.037
.038
.039
.040

Max. .041
Comp. .042
"A" .043
.044
.045
.046
.047

Max. .048
Comp.
"B"

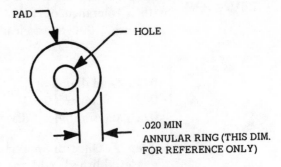

.020 MIN
ANNULAR RING (THIS DIM. FOR REFERENCE ONLY)

.037	HOLE DIAMETER (MAXIMUM)
+.040	.020 MINIMUM ANNULAR RING X 2
.077	MINIMUM SIZE PAD THAT COULD BE USED

OD		ID		CAT. NO.
in.	mm	in.	mm	
.050	1,27	.015	0,38	D135
.062	1,57	SOLID		S101
		.025	0,63	D136
.075	1,90	SOLID		S121
.080	2,03	.031	0,79	D216
.088	2,23	SOLID		S119
.093	2,36	SOLID		S102
		.031	0,79	D137
.100	2,54	SOLID		S123
		.031	0,79	D101
		.040	1,02	D207
		.050	1,27	D211
		.062	1,57	D148
.125	3,17	SOLID		S103
		.025	0,63	D180
		.031	0,79	D102
		.040	1,02	D179
		.050	1,27	D230
		.062	1,57	D169
.150	3,81	.031	0,79	D144
		.040	1,02	D203
		.050	1,27	D164
		.062	1,57	D231
		.080	2,03	D165

TABLE 4-1

TYPICAL PAD SIZE SELECTION CHART

To choose a size that will fit both component "A" and "B", select a size above the minimum of "A" and "B" but not to exceed the maximum of "A" and "B"; .037-in. or .038-in. could be used. To be exact, .035-in. through .040-in. diameters.

In conclusion of Step 4, the hole size for Figure 4-5 component would be between the minimum of .027-in. diameter and the maximum of .041-in. diameter depending upon the other components on our P.C. board such as the above.

Step 5—At this point, a hole size for the component or components could be established. Next a pad size for the P.C. board can be selected using the formula previously given in this chapter.

Pad Size Formula: .020-in. minimum annular ring over the maximum hole size. (.020 min. annular ring is for reference only) Use .037-in. diameter for hole size.

| OD | | ID | | CAT. |
in.	mm	in.	mm	NO.
.156	3,96	SOLID		S104
		.031	0,79	D103
.175	4,45	SOLID		S120
.187	4,75	SOLID		S122
		.025	0,63	D181
		.031	0,79	D138
		.040	1,02	D104
		.050	1,27	D150
		.062	1,57	D105
		.093	2,36	D182
.200	5,08	SOLID		S124
		.025	0,63	D183
		.031	0,79	D139
		.040	1,02	D204
		.050	1,27	D158
		.062	1,57	D147
		.062	1,57	D206
		.080	2,03	D106
		.093	2,36	D184
.208	5,28	.062	1,57	D168
.218	5,54	SOLID		S105
		.031	0,79	D140
		.040	1,02	D143
		.050	1,27	D155
		.062	1,57	D107

Step 6—The next step and last is to choose a pad that will fit or accommodate as many hole sizes as possible. Try not to select too many different sizes. Table 4-1 is a typical pad size selection chart.

Using the minimum size pad established in Step 5 (which is .077-in. diameter), convert all our diameters and pad diameters to the scale that the layout and tape-up are made (2:1 scale). Therefore, .077-in. x 2 = .154-in. diameter minimum. Now look at the pad size chart in Table 4-1 and select a pad size above .154-in. diameter.

NOTE: A good choice would be .187-in. diameter on the chart, but .156-in. through .350-in. could be used. The most important thing to remember is not to come too close to the minimum size that could be used, if possible.

A scale of 2/1 was used as the example of Step 6; a scale of 4/1 could have been used.

Chapter

5

LOGIC

The printed circuit board design process is evolutionary and consists of many phases. These phases may or may not be accomplished in a specific order. All however, will ultimately be required and accomplished through necessity. Digital logic is one of the evolutionary phases. Since you may not even know the meaning of the words "digital logic," it is the primary intent of this chapter to describe digital logic as an aid to the printed circuit board design process.

Component combining and grouping which are the real secrets to a successful PCB layout is a secondary intent of this chapter. Grouping will be discussed following the digital logic discussion. It is felt that an understanding of digital logic is a must before component grouping can be accomplished successfully.

This chapter will not make you an electronics technician nor will it make you an electronics engineer. Instead, you will have sufficient knowledge of digital logic to perform your P.C. board designing tasks more effectively and efficiently.

Experience has shown that understanding digital logic and applying this knowledge can solve many problems in the initial design stages rather than at the scheduled completion time. It is through problem solving that an understanding of digital logic is without a doubt an aid in the P.C. board design process.

The word "digital" is derived from the word digit which means number, such as the 1, 2, 3, 7, 9 digits. However, in the science of modern electronics certain words, or jargon have taken on a different meaning from those published by Webster. Also, certain other words are assumed to be included in many electronic definitions. For instance, the word digital not only means digits, but implies "binary," which means two-state (as opposed to decimal which means ten-state). Therefore, combining the apparent and not-so-apparent definition, the word digital, as you will know it, means: digits that can be in two states (two-state electronics). These two states are: 0, *Off* or Low and 1, *On*, or High.

Logic is the science of valid, efficient reasoning. In other words, it is a way of performing a given task in the most efficient manner available to you. As in the definition of digital, the word logic implies digital electronics. Combining the apparent and not-so-apparent, the words "digital logic" mean: An effective decision making, method using two-state electronics. Digital integrated circuits or IC's, are the hardware through which the digital logic concept is made possible.

The IC is a relatively small, sealed package with pins protruding from the sides, ends, top, or bottom. The package types are; dual-in-line packages (DIP) (See Figure 5-1) flat pack, and T0-5 can. As the name implies, *circuits* NOT just *components* are integrated, built-into, the IC. There are four common families of Integrated Circuits available today. They are:

SSI - Small Scale Integration
MSI - Medium Scale Integration
LSI - Large Scale Integration
VLSI - Very Large Scale Integration

The difference between these four is not the types of circuits contained within each package, but the density of circuitry contained within. Each of these four families can be further subdivided into the type of circuitry used to fabricate the particular IC device. These subdivisions are:

TTL - Transistor transistor logic
ECL - Emitter coupled logic
CMOS - Complementary metal oxide
        semiconductor logic
DTL - Diode transistor logic
RTL - Resistor-transistor logic
HTL - High-threshold logic

Our intention here is to point out some of the abbreviations used in the industry today, not to define each one in detail. The difference between these types are: switching speed, power consumption, cost, isolation, and sensitivity.

The word "circuit" warrants definition. A circuit, as opposed to a component, is made up of many discrete components such as resistors, capacitors, transistors, and diodes that can perform some function. For example, it might count up, count down, multiply or divide. In fact, it might do all at the same time. Whereas, the discrete component cannot perform a function readily without first being connected into a circuit.

Integrated circuits are usually formed on a silicon wafer consisting of a multitude of circuits. Various photographic and chemical etching processes are performed to produce the pre-designed electronic circuits. Then the individual circuits are separated from the wafer and are usually called die or chips. Gold wires are connected

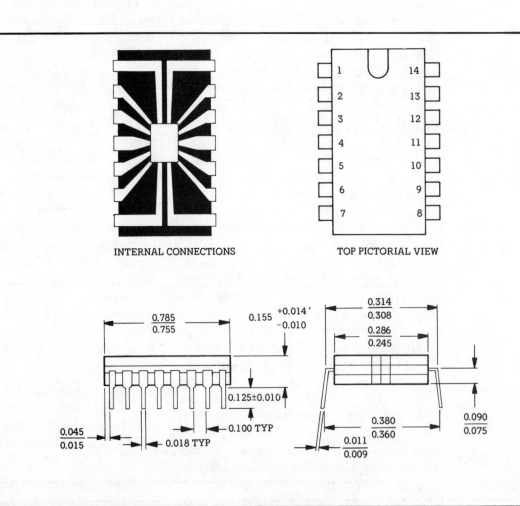

INTERNAL CONNECTIONS   TOP PICTORIAL VIEW

FIGURE 5-1
Dual Inline
Package (DIP)

PLASTIC DUAL IN-LINE PACKAGE TYPE P

16-LEAD PLASTIC DUAL IN-LINE PACKAGE TYPE P

.835 (21.209)
.825 (20.955)

PIN 1

.255 (6.477)
.245 (6.223)

.200 (5.080) MAX.

.140 (3.556)
.130 (3.302)

SEATING PLANE

.125 (3.175) MIN.

.015 MIN. (0.381)

.110 (2.794)
.090 (2.286)

.060 TYP (1.524)

.032 TYP (0.813)

.020 (0.508)
.016 (0.406)

.325 MAX. (8.255)

.010 TYP (0.254)

0° / 15

.350 (8.890) REF.

LEADS INSERT ON .100" x .300" CENTERS

18-LEAD PLASTIC DUAL IN-LINE PACKAGE TYPE P

.905 (22.987)
.895 (22.733)

PIN 1

.255 (6.467)
.245 (6.223)

.200 (5.080) MAX.

.140 (3.556)
.130 (3.302)

SEATING PLANE

.125 (3.175) MIN.

.015 MIN. (0.381)

.110 (2.794)
.090 (2.286)

.060 TYP (1.524)

.032 TYP (0.813)

.020 (0.508)
.016 (0.406)

.325 MAX. (8.255)

.010 TYP (0.254)

0° / 15

.350 (8.890) REF.

LEADS INSERT ON .100" x .300" CENTERS

20-LEAD PLASTIC DUAL IN-LINE PACKAGE TYPE P

1.035 (26.289)
1.025 (26.035)

PIN 1

.255 (6.467)
.245 (6.223)

.200 (5.080) MAX.

.140 (3.556)
.130 (3.302)

SEATING PLANE

.125 (3.175) MIN.

.015 MIN. (0.381)

.110 (2.794)
.090 (2.286)

.060 TYP (1.524)

.032 TYP (0.813)

.020 (0.508)
.016 (0.406)

.325 MAX. (8.255)

.010 TYP (0.254)

0° / 15

.350 (8.890) REF.

LEADS INSERT ON .100" x .300" CENTERS

22-LEAD PLASTIC DUAL IN-LINE PACKAGE TYPE P

1.105 (28.067)
1.095 (27.813)

PIN 1

.355 (9.017)
.345 (8.763)

.200 (5.080) MAX.

.155 (3.937)
.145 (3.683)

SEATING PLANE

.125 (3.175) MIN.

.015 MIN. (0.381)

.110 (2.794)
.090 (2.286)

.060 TYP (1.524)

.032 TYP (0.813)

.020 (0.508)
.016 (0.406)

.425 MAX. (10.795)

.010 TYP (0.254)

0° / 15°

.450 (11.430) REF.

LEADS INSERT ON .100" x .400" CENTERS

FIGURE 5-1 (cont.)

LOGIC

69

PLASTIC DUAL IN-LINE PACKAGE TYPE P

24-LEAD PLASTIC DUAL IN-LINE PACKAGE TYPE P

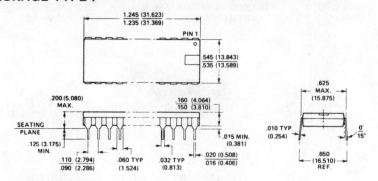

LEADS INSERT ON .100" x .600" CENTERS

28-LEAD PLASTIC DUAL IN-LINE PACKAGE TYPE P

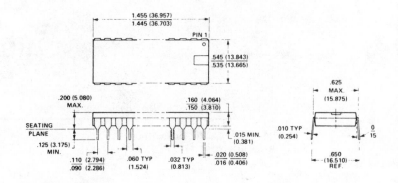

LEADS INSERT ON .100" x .600" CENTERS

40-LEAD PLASTIC DUAL IN-LINE PACKAGE TYPE P

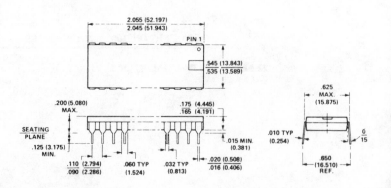

LEADS INSERT ON .100" x .600" CENTERS

FIGURE 5-1
(cont.)

LOGIC

between the output points or the chip and I.C. pins. Then the I.C. is sealed. This process is the same for all I.C.'s made today. (SSI, MSI, LSI, and VLSI).

Every company that uses IC's in their electronic designs can choose from standard or custom devices. Standard devices are relatively inexpensive and perform some fairly standard functions. Whereas custom devices are specially designed for a specific application and are generally more expensive. Examples of fairly standard IC's are the ones contained in calculators available today or the IC radio you've listened to. An example of a custom device is the IC used in the more scientific programmable hand-held calculators, sophisticated computer circuits, and in some of the electronic TV tuners now available. Both standard and custom devices are used in today's computer technology, as well as in our automobiles and home entertainment equipment.

IC's are by no means magic. They perform as a function of what has been put inside. Today, there are literally thousands of IC types produced by many IC manufacturers. Don't get worried, each IC pack has an identifying sequence of standardized numbers and most manufacturers comply with accepted numbering rules.

Data sheets describe what is inside the chip. All manufacturing companies, especially the larger ones, produce exceptionally good data sheets for their IC devices.

IC's have created a new, more efficient method of designing electronic circuits. Circuits designed using IC's also require the use of symbols to define the circuit to its reader: First for the electronic engineer, second for the PCB designer, then as a tool for the maintenance people to use in the field. Digital logic symbols are, for the most part, unlike those used in the old vacuum tube days. Some symbols used today haven't changed much in appearance and include discrete components used in modern analog circuits such as resistors and capacitors.

Digital logic symbology is unique in that each symbol is itself a condensed map of components. We refer to each symbol as a "gate." Gates perform specific functions within the IC and as mentioned consist of many components (resistors, capacitors, diodes, transistors) previously fabricated onto the IC chip. In most cases, an IC includes many gates. Diagrammed below is the semiconductor integrated circuit building block (See Figure 5-2.)

LOGIC RULES

Basic digital logic layouts use digital symbology; these layouts will be called "schematics." An understanding of digital logic symbols is extremely important to successful layout of digital P.C. Boards.

Basic logic rules apply to gates. Like your lights at home, each lamp has two basic states; *ON* and *OFF*. (A dimmed lamp is considered ON). To *activate* a lamp the appropriate switch is turned on. Likewise to *deactivate* the lamp the procedure is reversed. In the home, 115 volts of alternating current (AC) is used to activate the lamp. When the lamp is deactivated, voltage is removed and 0 volts of alternating current are supplied. Digital logic gates also have two states and are activated and deactivated by the application and removal of a voltage. But that's where the likeness stops. Logic gates require direct current (DC) signals and in most cases require more than a single action to function. Unlike the light switch above, all that was necessary was one *input* (a flick of the switch lever) to get one output (the light ON or OFF), while each logic gate can require several *inputs* and usually provides one *output*.

For this discussion, an *ON* or Logic 1

state is identified as +5 Vdc and an *OFF* or Logic 0 state as 0 Vdc. 0 Vdc is actually ground potential and is abbreviated GND. This arrangement of +5 Vdc = Logic 1 and 0 Vdc = Logic 0 is referred to as positive true logic, meaning the more positive voltage is a Logic 1. Negative true logic could be used where the less positive voltage is a Logic 1 and the +5 Vdc and 0 Vdc levels could be other values. These rules can be applied to an actual digital logic gate symbol. Figure 5-3 below is what we call the *AND* gate.

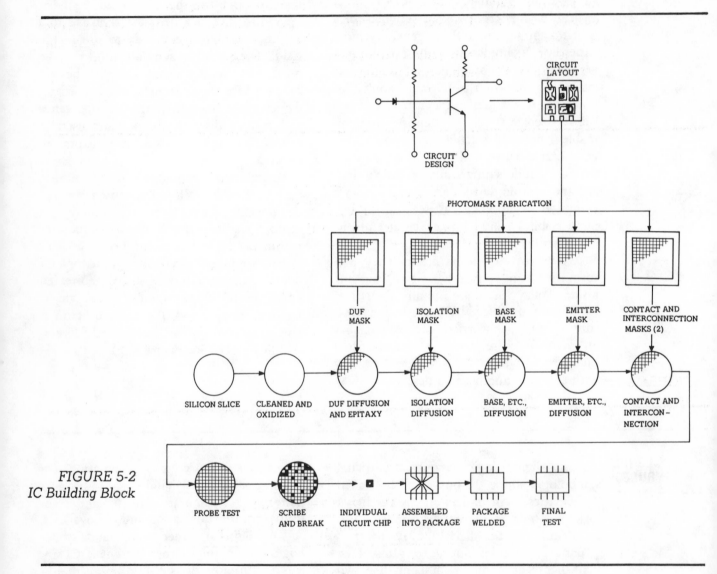

FIGURE 5-2
IC Building Block

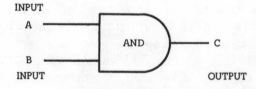

FIGURE 5-3
Basic Logic Gate

LOGIC

Leads A & B are *inputs,* Lead C is an *output.* There could have been many more inputs (designated as the fan-in), however, for this explanation only two are used. To make this gate function, we want the output (C) to change state when we make certain changes to the input leads (A and B). Now remember, only two states exist on any one lead whether it be input or output. Figure 5-4 describes the operation of Figure 5-3.

Most IC data sheets provide a truth table which shows all possible combinations of input conditions and the resulting output. For instance, the different conditions that exist for two input leads are shown in Figure 5-5.

As can be seen, a two-input gate can have four completely different input configurations in two-state electronics. With three inputs, there are eight configurations or states, see Figure 5-6.

As mentioned, the data sheet truth table will also provide output states from the various inputs.

STATE	INPUTS		OUTPUT
	A	B	C
1	0	0	0
2	0	1	0
3	1	0	0
4	1	1	1

Where 1 = +5Vdc or Logic 1
0 = 0Vdc or Logic 0

FIGURE 5-4

TWO INPUT TRUTH TABLE

STATES	INPUT A	INPUT B
1	Logic 0	0
2	Logic 0	1
3	Logic 1	0
4	Logic 1	1

FIGURE 5-5

THREE INPUT TRUTH TABLE

STATE	INPUT A	INPUT B	INPUT C
1	0	0	0
2	0	0	1
3	0	1	0
4	0	1	1
5	1	0	0
6	1	0	1
7	1	1	0
8	1	1	1

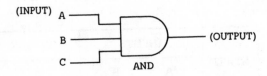

FIGURE 5-6

DIGITAL LOGIC SYSTEMS

Digital logic symbols fall into two categories: basic gate symbols and complex gate symbols. There are three gate symbols in the basic category. These are:

- AND Gate
- OR Gate
- INVERTER

The complex gate symbols include many variations. The symbol used to represent these variations is a rectangular box. Because there are standard and special devices that fall into the complex gate category, all that will be discussed here are the standard flip flop devices. These are:

- Basic flip-flop
  (Commonly known as the RS Flip-flop or latch)
- JK flip-flop
- T flip-flop
- D flip-flop

Special devices that will not be discussed in this chapter include:

- Shift registers
- Counters
- One-shot multivibrator

These devices also use the rectangular box as the logic symbol.

Actually, all complex devices;

- Flip-flop
- Shift registers
- Counters
- One-shot multivibrator

are composed of the three basic gates; AND, OR, and INVERTER even though some data sheets don't show them this way. However, flip-flop packs are widely used to understand their basic operation.

Before we attempt a discussion of logic gate generation, let's review some basic rules:

- There are 3 basic logic symbols; AND, OR, INVERTER.
- All logic gates have at least one input lead.
- Most logic gates have a single output lead.
- The normal Positive Logic convention is:

    An *ON* state is a Logic 1, or High.
    An *OFF* state is a Logic 0, or Low.

- The NAND gate (not AND) is the AND gate with an inverter on the output lead.
- The NOR gate (not OR) is the OR gate with an inverter on the output lead.
- Logic symbols represent circuits within an IC package.
- Each IC package contains at least one logic symbol (hence circuit) usually more.
- The total number of inputs on a gate refers to fan-in.

BASIC LOGIC SYMBOL DEFINI-TIONS

The AND gate logic symbol is shown in fig. 5-7 along with a dual inline (DIP) IC that includes four 2 input AND gates (called a Quad 2 input AND).

The AND gate shown in Figure 5-7 is a device whose output pin is a Logic 1 when, and only when, all of the inputs are also a Logic 1 as shown in State 4 of the AND Truth Table. When any input goes to a Logic 0 state, the AND gate output also goes to a Logic 0 as shown in States 1-3 of the AND Truth Table. The AND gate can include up to 8 input leads. If only one input is ever used, the device no longer functions as an AND gate. Rather it functions as a buffer element. An analogy to the AND gate is a lamp with two (or more) switches required to turn the bulb *ON*. In Figure 5-8 below, Switch 1 and Switch 2 must be *ON* to energize the light bulb.

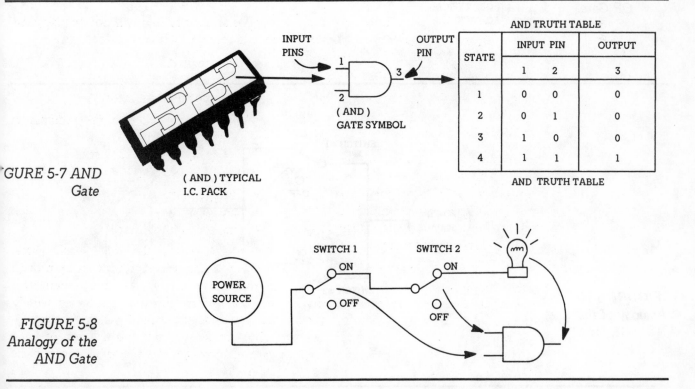

AND TRUTH TABLE

STATE	INPUT PIN		OUTPUT
	1	2	3
1	0	0	0
2	0	1	0
3	1	0	0
4	1	1	1

AND TRUTH TABLE

GURE 5-7 AND Gate

(AND) TYPICAL I.C. PACK

FIGURE 5-8 Analogy of the AND Gate

If either or both Switch 1 and 2 are OFF, the light bulb will also be OFF. Therefore, the AND gate is actually a series element. That is, it requires all input states to be *ON* before the output state can be *ON*.

The symbol for a 2-input OR gate is shown in Fig. 5-9, along with a Quad 2-input, DIP or Gate.

The OR gate shown in Figure 5-9 is a device whose output pin is a Logic 1 when any or all inputs are a Logic 1, as shown in states 2-4 in the Truth Table. When, and only when, all inputs are a Logic 0, as shown in state 1 of the Truth Table, will the OR gate output be a Logic 0. The OR gate can include up to ten input leads. If only one input is ever used, the device no longer functions as an OR gate. Rather it functions as a buffer element. An analogy to the OR gate is a lamp with two (or more) switches, where *each* switch can supply power directly to the lamp and light it. In Figure 5-10 below, Switch 1 or Switch 2 or both switches 1 and 2 being on will energize the bulb. The lamp will only go OFF when both switches are OFF.

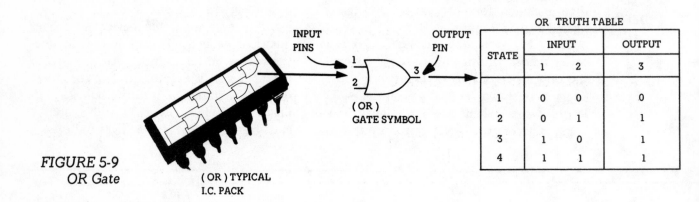

FIGURE 5-9
OR Gate

INPUT PINS

OUTPUT PIN

(OR)
GATE SYMBOL

(OR) TYPICAL
I.C. PACK

STATE	INPUT		OUTPUT
	1	2	3
1	0	0	0
2	0	1	1
3	1	0	1
4	1	1	1

OR TRUTH TABLE

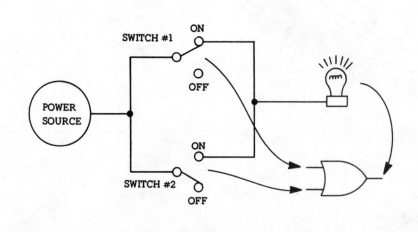

FIGURE 5-10
Analogy of the OR Gate

SWITCH #1
ON
OFF

POWER SOURCE

ON
SWITCH #2
OFF

Therefore, the OR gate is actually a parallel element. That is, it can be in the *ON* state by any input or inputs or all inputs being in the *ON* state.

The inverter of Fig. 5-11 inverts an incoming signal. Also shown in the Figure is a DIP, Hex inverter (six inverters).

The inverter shown in Figure 5-11 is a device whose output is a Logic 0 when the input is a Logic 1 as shown in State 2 of the inverter Truth Table. The inverter output is a Logic 1 when the input is a Logic 0 as shown in State 1 of the Truth Table. NAND gates and NOR gates as shown in Figure 5-12 below, with all input leads tied together also function as inverters.

An analogy to the inverter requires some reverse thinking. For example, when you turn your lamp switch *ON* you expect the lamp to go ON. If, however, the switch was wired backward the lamp would go OFF when the switch was turned on. The inverter functions this way, when the output state is reversed from the input state.

The Nand gate symbol of Fig. 13 is similar to that of the And gate with a small circle at its output terminal.

The NAND gate shown in Figure 5-13 is actually an AND gate with a "built-in" inverter on the output lead as shown just below NAND GATE Symbol above. The device output pin, as shown in States 1-3 of the NAND Truth Table, is a Logic 1 when any or all inputs are a Logic 0. The NAND gate output goes to a Logic 0 state when and only when *all* inputs are a Logic 1, State 4 on the Truth Table.

An analogy to the NAND gate, like the inverter, requires reverse thinking. An *ON* lamp is normally associated with pushing a button in or flipping a switch *UP*. In Figure 5-14 below, the lamp is always *ON*, powered directly from a power source. Switches 1 and 2 control the power source. As long as Switches 1 and 2 or both are OFF, the power source is not grounded out and it continues to power the lamp. If, however, both switches go ON, the power source is grounded out (the power source is turned OFF) and the lamp goes OFF.

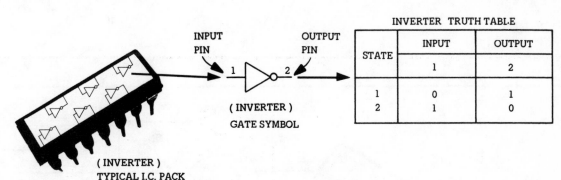

INVERTER TRUTH TABLE

STATE	INPUT	OUTPUT
	1	2
1	0	1
2	1	0

INPUT PIN   OUTPUT PIN

1 — 2

(INVERTER)
GATE SYMBOL

(INVERTER)
TYPICAL I.C. PACK

FIGURE 5-11
Inverter Gate

FIGURE 5-12
NAND and OR
tes Connected
as Inverters

Therefore, the NAND gate is a serial element just like the AND gate except that the output is inverted.

The NOR gate Symbol of Figure 5-15 is similar to the *OR* gate with a small circle at the output.

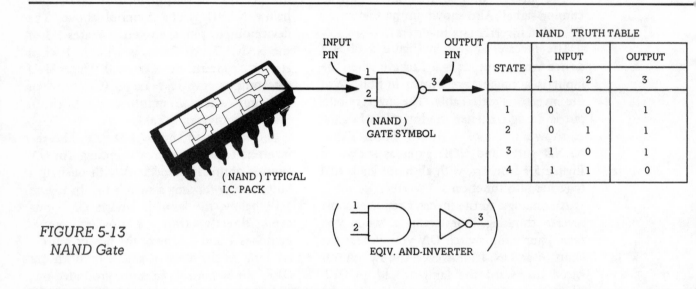

INPUT PIN   OUTPUT PIN

(NAND) GATE SYMBOL

(NAND) TYPICAL I.C. PACK

EQIV. AND-INVERTER

STATE	NAND TRUTH TABLE		
	INPUT		OUTPUT
	1	2	3
1	0	0	1
2	0	1	1
3	1	0	1
4	1	1	0

FIGURE 5-13
NAND Gate

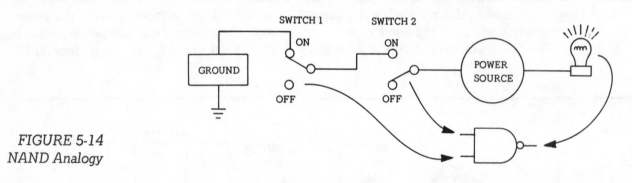

SWITCH 1   SWITCH 2

ON   ON

GROUND

OFF   OFF

POWER SOURCE

FIGURE 5-14
NAND Analogy

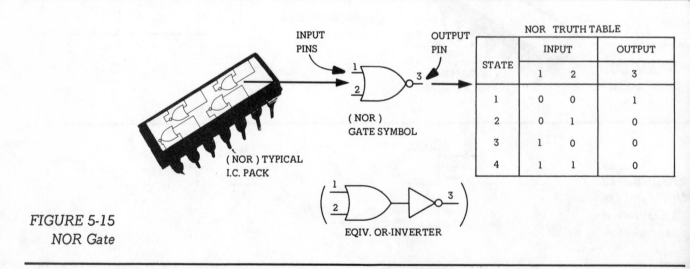

INPUT PINS   OUTPUT PIN

(NOR) GATE SYMBOL

(NOR) TYPICAL I.C. PACK

EQIV. OR-INVERTER

STATE	NOR TRUTH TABLE		
	INPUT		OUTPUT
	1	2	3
1	0	0	1
2	0	1	0
3	1	0	0
4	1	1	0

FIGURE 5-15
NOR Gate

LOGIC

The NOR gate shown in Figure 5-15 above, is actually an OR gate with a "built-in" inverter on the output lead, as shown just below NOR gate symbol above. The device output pin, as shown in State 1 of the NOR Truth Table, is logically a Logic 1 only when all inputs are in Logic 0 state. The NOR gate output pin is a Logic 0 when any or all inputs are in a Logic 1 state, as shown in States 2-4.

An analogy to the NOR gate, like the inverter, and NAND gate requires reverse thinking. As with the NAND gate analogy, an *ON* lamp is associated with pushing a button *in* or flipping a switch up. In Figure 5-16 below, the lamp is always *ON*, powered directly from a power source. The

power source is controlled by Switches 1 and 2. If a ground is supplied to the power source, the lamp will go OFF. As can be seen, both Switch 1 and 2 are connected directly to a ground. If either switch individually or both are turned *ON*, the power source is grounded out (turned-off) and the lamp goes off. Therefore, the NOR gate is a parallel element just like the OR gate except that the output is inverted.

The Exclusive-OR Gate shown in Figure 5-17 is similar to the OR Gate. The Exclusive-OR Gate output, however, is a Logic 1 when and only when *one* of the inputs is in a Logic 1 state, as shown in States 2 and 3 in the Truth Table below. When all inputs are either a Logic 0 or a Logic 1, the

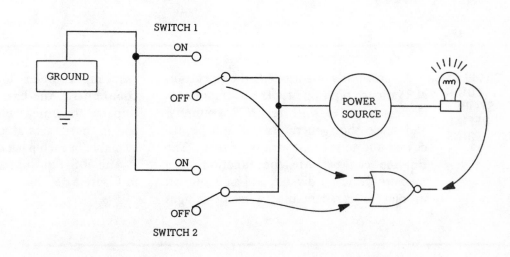

FIGURE 5-16
NOR Analogy

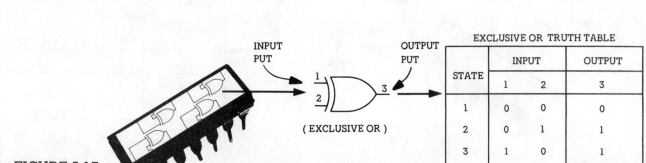

FIGURE 5-17
xclusive OR Gate

	EXCLUSIVE OR TRUTH TABLE		
	INPUT		OUTPUT
STATE	1	2	3
1	0	0	0
2	0	1	1
3	1	0	1
4	1	1	0

THIS IS NOT A STANDARD PRACTICE, BUT MAY BE USED TO IDENTIFY OPEN COLLECTOR GATES

OPEN-COLLECTOR NAND GATE   OPEN-COLLECTOR INVERTER

FIGURE 5-18
Open Collector

Exclusive-OR gate output is a Logic 0 as shown in States 1 and 4 above. The Exclusive-OR gate is therefore a type of lock-out device. It locks out all Logic 0 states at input as well as more than one Logic 1 state.

The logic gates shown in Figure 5-18 above are of the open-collector type. These gates are used in timing circuits, as drivers and for wired-AND functions. If a logic gate is connected to the output of the open-collector gate, it then functions just like a normal NAND gate or inverter. The manufacturing part number is the only means of identifying a open-collector gate from a non-open collector gate.

COMPLEX LOGIC SYMBOL DEFINI-TIONS

As previously mentioned, the operation of various flip-flop types is important. Devices such as shift registers, counters, and single shot generators will not be discussed due to the vast variety of them. The flip-flop can perform the function of a memory device, a divider, and a counter all in one. It is adaptable to many designs requiring complex logic devices. Like its counterparts, the basic logic gates, the flip-flop is contained within an IC package and in most cases at least two flip flops are available in each package.

The R-S flip-flop is shown symbolically in Figure 5-19.

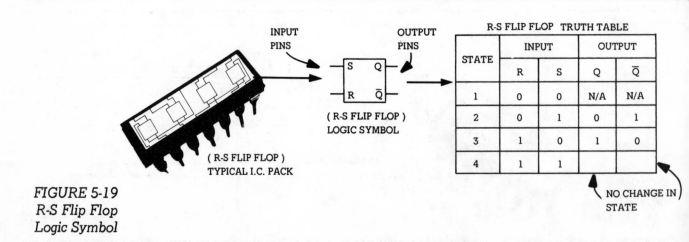

INPUT PINS

OUTPUT PINS

(R-S FLIP FLOP)
TYPICAL I.C. PACK

(R-S FLIP FLOP)
LOGIC SYMBOL

R-S FLIP FLOP TRUTH TABLE

STATE	INPUT		OUTPUT	
	R	S	Q	$\overline{Q}$
1	0	0	N/A	N/A
2	0	1	0	1
3	1	0	1	0
4	1	1		

NO CHANGE IN STATE

FIGURE 5-19
R-S Flip Flop
Logic Symbol

The R-S (Reset-Set) flip-flop is the most basic of flip-flops and is generally referred to as a latch. Actually, most latches are a combination of two cross-connected NAND gates, as shown in Figure 5-20 below.

As shown in the R-S Flip-Flop Truth Table, when a Logic 1 is applied to the S input lead (State 2) the $\bar{Q}$ output lead goes to a Logic 1 and Q goes to a Logic 0. When the R input lead goes to a Logic 1 (State 3) the flip/flop changes state; $\bar{Q}$ goes to a Logic 0 and Q goes to a Logic 1. When both R and S inputs are in a Logic 2 state, there is no change in the flip-flop's output state. A Logic 0 state on both input leads is not allowed. Q and $\bar{Q}$ are never in the same state.

Another form of flip-flop is the "D" type shown in Figure 5-21.

The "T" (toggle) or "clocked" flip-flop changes state as a result of an input pulse called a trigger, clock pulse or toggle at the T lead. As shown in the Truth Table, state 1, after the first toggle occurs, the Q and $\bar{Q}$ states change. When the second toggle occurs, the flip/flop reverts back to the original state with Q a Logic 0 and $\bar{Q}$ a Logic 1. If a trigger is placed on the SD lead, the flip-flop goes to State 1. If a trigger is placed on the RD lead, the flip-flop goes to State 2. As long as trigger pulses occur at the "T" input the flip-flop will change state back and forth. Q and $\bar{Q}$ states are never the same.

The J-K flip-flop shown in Figure 5-22 above is an important element in most digital electronic designs. This flip-flop offers complete flexibility in that all possible controls are "built-in" but not all are required for the gate to function. For example, the SD and RD (Set direct and reset Direct)

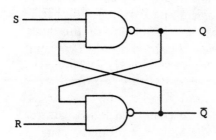

FIGURE 5-20
chematic of R-S
Flip Flop

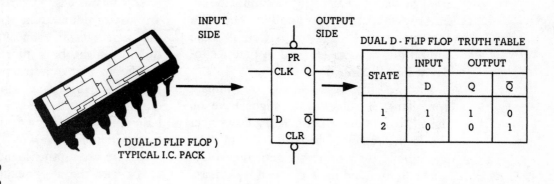

FIGURE 5-21
Dual - D Flip
Flop

(DUAL-D FLIP FLOP)
TYPICAL I.C. PACK

DUAL D - FLIP FLOP TRUTH TABLE

STATE	INPUT	OUTPUT	
	D	Q	$\bar{Q}$
1	1	1	0
2	0	0	1

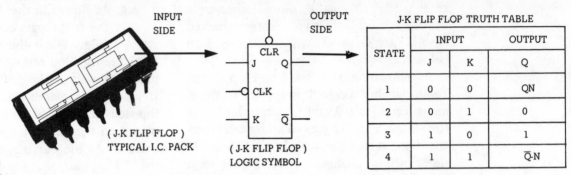

FIGURE 5-22
J-K Flip Flop

J-K FLIP FLOP TRUTH TABLE

STATE	INPUT		OUTPUT
	J	K	Q
1	0	0	QN
2	0	1	0
3	1	0	1
4	1	1	$\overline{Q}$-N

leads need never be used. In addition, by tieing the J and K input leads to a Logic 1 state, the flip-flop acts just like a "T" flip-flop already discussed.

As shown in the Truth Table, the J-K flip-flop Q output goes to a Logic 1 when the J input is a Logic 1, as in State 3 of the Truth Table, and a trigger occurs at T. The flip-flop $\overline{Q}$ output goes to a Logic 1, as in State 2 of the Truth Table, when the K input is a Logic 1 and a trigger occurs at T. Therefore, a Logic 1 on one of the input leads while a trigger pulse is occurring

causes the J-K flip-flop output to change state. The J-K flip-flop may have direct set (SD) and direct reset (RD) inputs, in which case a transition from a Logic 1 to a Logic 0 on either lead will cause the flip-flop to attain a certain state. When SD is triggered, the flip-flop reverts to a set state with Q a Logic 1, as in State 3 as the Truth Table. When RD is triggered, the flip-flop reverts to a reset state with $\overline{Q}$ a Logic 1, as in State 2 of the Table. Q and $\overline{Q}$ are never in the same state.

COMBINING   *Combining* is a term used by the author to describe methods for properly using a logic gate in applications other than what it was designed for. (Substitution might be a more appropriate word). It should be realized, however, that one gate cannot be arbitrarily substituted for another. The rules for combining logic gates as well as some typical design methods used in practice today will be described.

Without first considering the impact combining or substituting might have on the fit, form, and function of your project (be it the repair of a drain pipe, bicycle, automobile, or a printed circuit board layout) you may end up with a real nightmare. On the other hand, *good* combining or substitution habits will undoubtedly result in

the most economical and compact project possible.

A classic example of effects produced by substitution could be; using a spark plug not rated for a car's engine. The negative effects in this case might be poor gas mileage and/or poor engine ignition. Of course the mere fact that an automobile is now running might be a more positive effect. Similar to this example the side effects of substituting logic gates will become obvious during the board test phase. If some very basic rules are followed however, the effects will be positive.

Before attempting logic gate combining, check with the design engineer *and* refer to the manufacturer's specifications. This practice will ensure that gate combining wasn't the cause for the circuit's unsuccessful operation. Be aware of the following characteristics when combining and substituting logic gates:

- Open-Collector Gates vs. Non-Open Collector Gates
- Input Pins and Output Pin Numbers are compatible
- Combining Logic Gates should be done under the Engineer's direction.

To be more specific, combining is a process involving the numbers of logic gate input leads. Before going on, consider this example: If a structure must have four windows minimum to meet your needs having a fifth window would most likely be okay. If it weren't, you could "board-up" the extra window much easier than rebuilding the complete structure. Similarly, if a digital circuit requires a two input NAND gate and a four input NAND gate is available, the spare input leads can be ignored so the gate functions as a two-input device. (Much better than adding a complete IC package). The whole point is to conserve space and save money by using what's available.

Some additional rules to remember follow:

AND - When combining AND Gates, the unused (spare) input lead(s) *MUST BE* connected to *Vcc or to a functioning input lead. (See Figure 5-23)

NAND - When combining NAND gates the unused (spare) input lead(s) *MUST BE* connected to *Vcc or to a functioning input lead. (See Figure 5-24)

OR - When combining OR Gates, the unused (spare) input lead(s) *MUST BE* connected to ground or to a functioning input lead. (See Figure 5-25)

EXCLUSIVE-OR - When combining Exclusive-OR Gates the unused input lead(s) *MUST BE* connected. (See Figure 5-26)

NOR - When combining NOR Gates the unused input lead(s) *MUST BE* connected to ground or to a functioning input lead. (See Figure 5-27)

EXCLUSIVE-NOR - When combining Exclusive-NOR Gates, the unused input lead(s) *MUST BE* connected. (See Figure 5-28)

*Vcc is the positive supply voltage applied

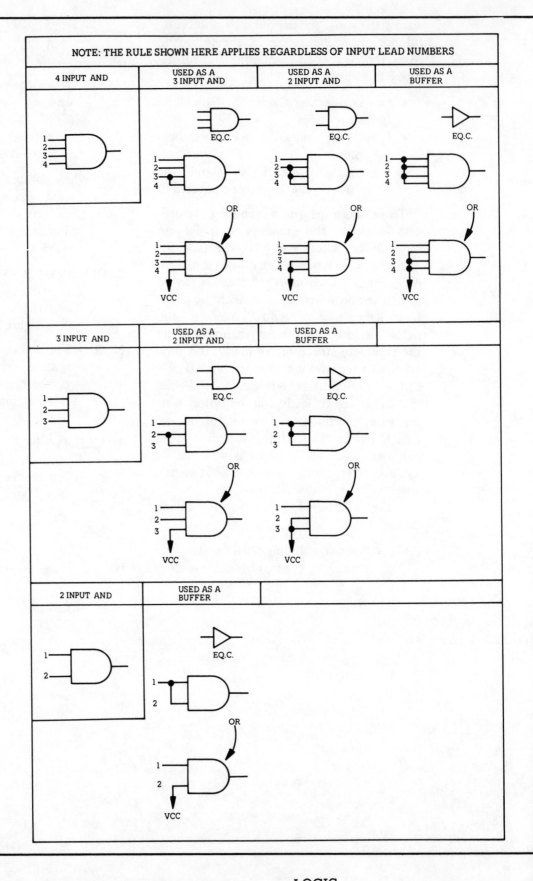

FIGURE 5-23
AND Gate
Combining

LOGIC

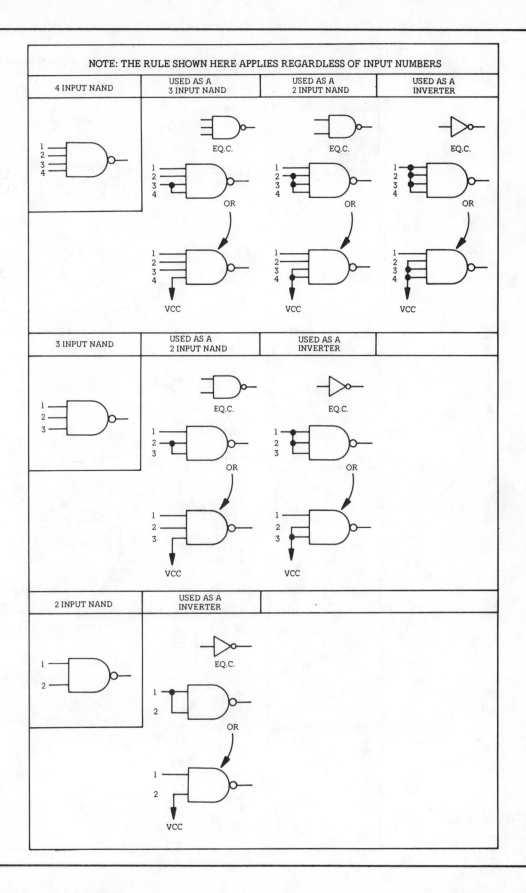

FIGURE 5-24
NAND Gate
Combining

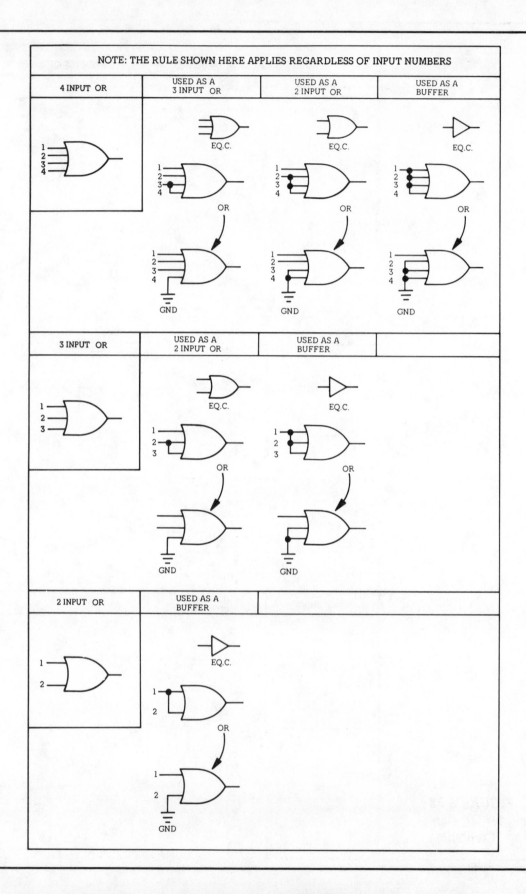

FIGURE 5-25 OR
Gate Combining

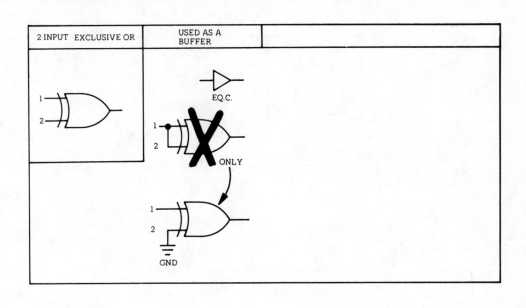

2 INPUT EXCLUSIVE OR	USED AS A BUFFER	

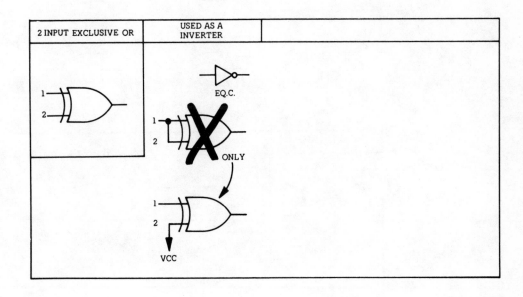

2 INPUT EXCLUSIVE OR	USED AS A INVERTER	

FIGURE 5-26
Exclusive OR Gate
Combining

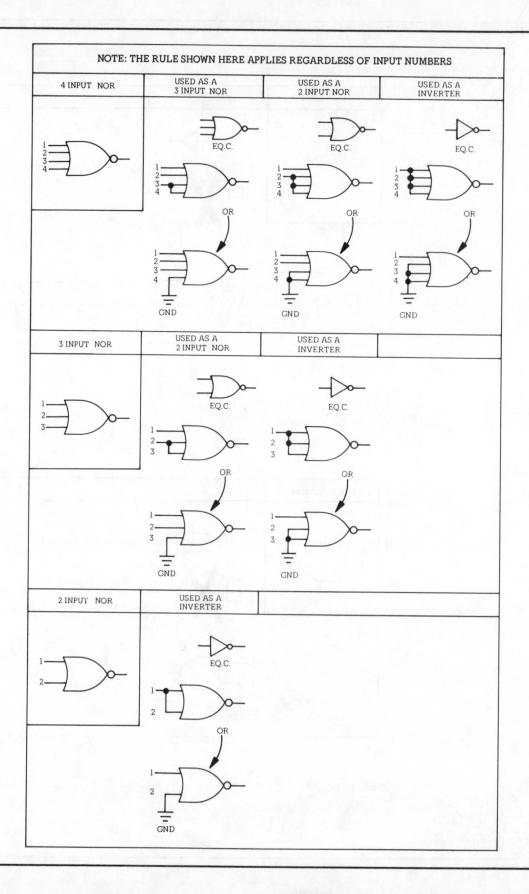

FIGURE 5-27
NOR Gate
Combining

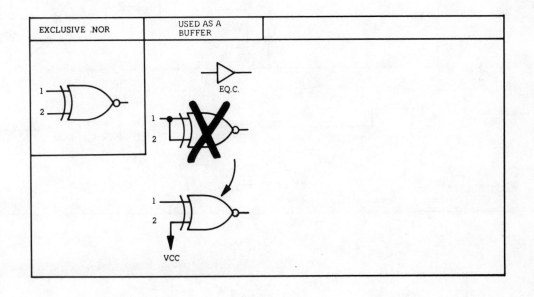

EXCLUSIVE .NOR	USED AS A BUFFER	

FIGURE 5-28
Exclusive NOR
Gate Combining

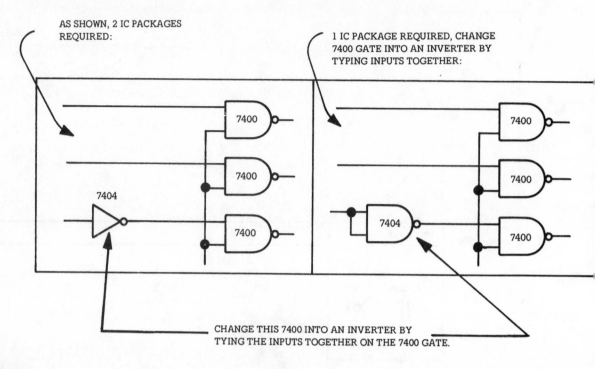

AS SHOWN, 2 IC PACKAGES REQUIRED:

1 IC PACKAGE REQUIRED, CHANGE 7400 GATE INTO AN INVERTER BY TYPING INPUTS TOGETHER:

CHANGE THIS 7400 INTO AN INVERTER BY TYING THE INPUTS TOGETHER ON THE 7400 GATE.

FIGURE 5-29
Combining

OLD

NEW

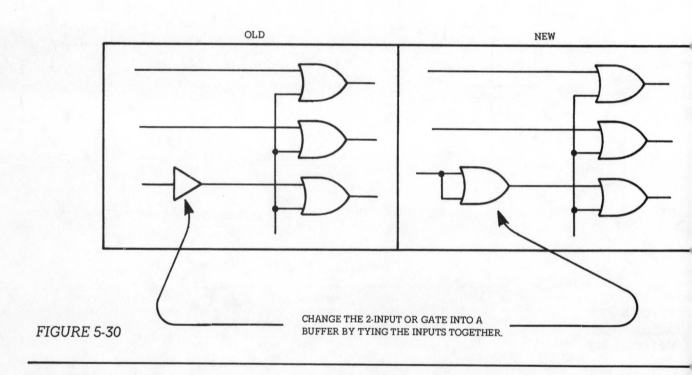

CHANGE THE 2-INPUT OR GATE INTO A BUFFER BY TYING THE INPUTS TOGETHER.

FIGURE 5-30

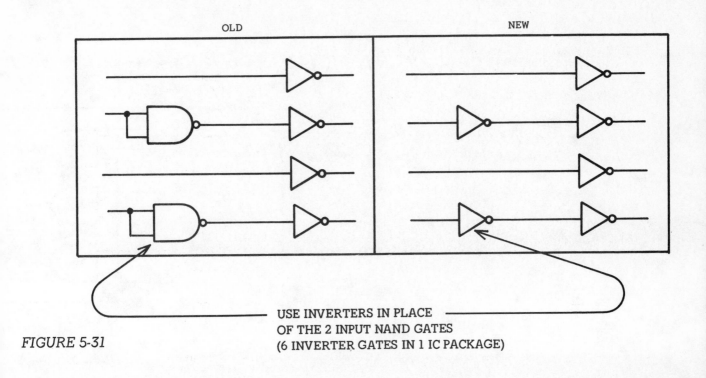

USE INVERTERS IN PLACE
OF THE 2 INPUT NAND GATES
(6 INVERTER GATES IN 1 IC PACKAGE)

FIGURE 5-31

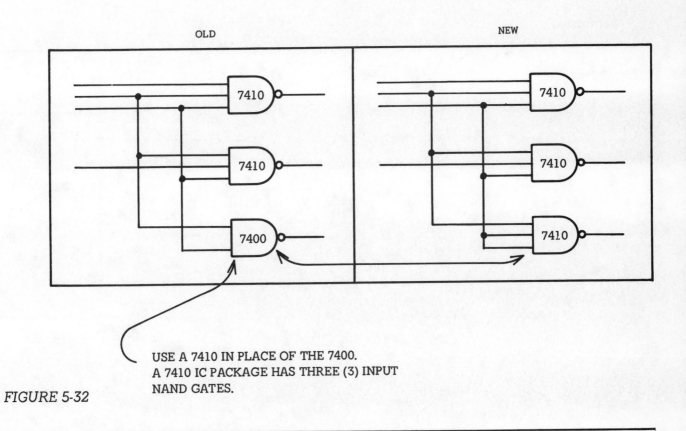

USE A 7410 IN PLACE OF THE 7400.
A 7410 IC PACKAGE HAS THREE (3) INPUT
NAND GATES.

FIGURE 5-32

Chapter

6

INTEGRATED CIRCUIT GROUPING

Grouping digital logic components is a process in which the total number of IC devices required in the design are identified. And once identified the gates are combined to use as many spares as possible, then space is allocated (on a drawing representing the PCB) for each integrated circuit package.

The system sketch, discussed in Chapter 1, could be used for a final PCB layout, however, because it is rough it is highly recommended that time be allotted for a schematic redraw. Using a rough schematic for the layout could result in problems, the most likely of which are listed below:

- Poor PCB design resulting in poor circuit operation.
- Excessive cost resulting from many unused logic gates (too many spares).
- Excessive PCB and Connector real estate required as a result of poor combining, grouping and layout practice.
- Excessive design time resulting from unsuccessful design attempts.
- Excessive time on the street looking for a new job.

It is important that IC grouping be accomplished before the PCB layout has begun. A discussion of this phase follows. It will become clear why grouping is done at the schematic phase and what specific problems can be expected by doing otherwise.

NOTE: Always verify combining of gates with the design engineer. As the PCB designer, you are a recommender. The design engineer should make the decisions.

Identifying the logic gates simply requires a reference to manufacturer specification sheets. This practice is usually done by the engineer, however, this work should be double checked. Reference to these specifications also provides the opportunity to identify the total number of logic gates in each package and the respective gate pin numbers. A table of each logic gate by IC package type should be generated and each gate checked off as it is assigned an actual location on the schematic. In this way, it will be apparent how many spares exist. Figure 6-1 shows a circuit and the IC's assigned by the engineer. The figure also shows a table of each IC package and how each is checked off once assigned to the schematic. Be particularly observant of the spare gates available in each IC package. At this time eliminate as many IC gates as possible. In the example, the two-input NOR IC-5 can be replaced by the two spares of the three input NOR IC-7. Before doing this check with the design engineer for complete compatability. The design engineers may want extra three-input Nor's on the board for future expansion.

NOTE: You have *NOT* assigned gate pin numbers at this point.

Once convinced the minimum number of IC packs are being used, proceed to assign gates to the schematic by area.

Figure 6-2 shows how to assign IC packs by area. This assigning process is done to keep traces short and helps eliminate the need for many PCB feed-throughs. This is where *grouping* plays its most important role. Successful grouping, as shown in Figure 6-2 places the "grouped" gates in close proximity to one another and to connector pins.

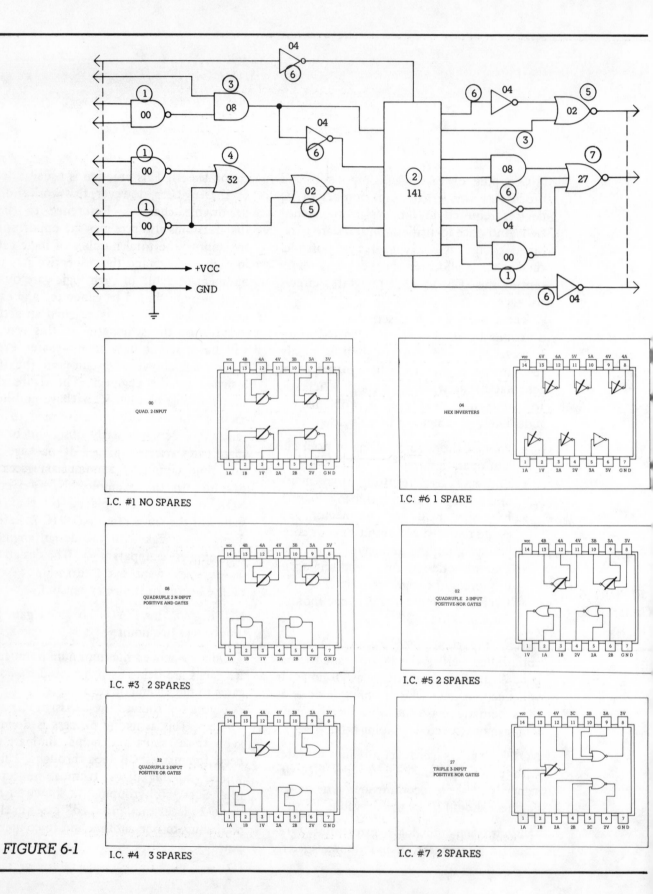

FIGURE 6-1

I.C. #1 NO SPARES

I.C. #6 1 SPARE

I.C. #3 2 SPARES

I.C. #5 2 SPARES

I.C. #4 3 SPARES

I.C. #7 2 SPARES

INTEGRATED CIRCUIT GROUPING

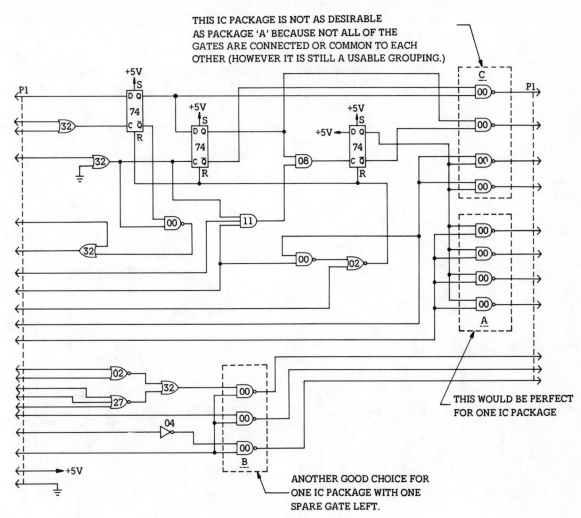

THIS IC PACKAGE IS NOT AS DESIRABLE
AS PACKAGE 'A' BECAUSE NOT ALL OF THE
GATES ARE CONNECTED OR COMMON TO EACH
OTHER (HOWEVER IT IS STILL A USABLE GROUPING.)

THIS WOULD BE PERFECT
FOR ONE IC PACKAGE

ANOTHER GOOD CHOICE FOR
ONE IC PACKAGE WITH ONE
SPARE GATE LEFT.

FIGURE 6-2
Grouping IC's

The rules still apply, regardless of the complexity.

- Determine gate requirements or numbers.
- Combine where possible (referring to manufacturer's specifications).
- Group gates by IC packs to enhance the mechanics of the upcoming design (keeping traces short, close to connectors, while minimizing feedthrough requirements).

Take a good *overall* look at the schematic. Smaller portions are already "grouped" to facilitate signal flow. This flow will determine how the actual PCB is to be laid out. For example, in Figure 6-3 the outputs from both IC 3 and 4 are connected directly into input to IC 2. Likewise, IC 5 outputs are connected to the inputs to IC 6. Furthermore, IC's 3, 4, and 5 inputs come directly from the connector P1. IC 1 and 2 outputs are applied to IC 7, IC 7 and 14 outputs are applied to the inputs of IC's 8, 9, 10, and 11 and so on. This is the flow in Figure 6-3. The schematic should flow in a similar manner.

Undoubtedly, gates will feed to the left, up, or down. These are exceptions, in most cases, and cause no problems in the designed layout, as long as *most* of the schematic is consistent with the proper flow. *Keep the flow in mind and don't be concerned about gates that don't fit the overall left-to-right flow pattern.*

Now that all gates are assigned and grouped look at their placement on the layout. First count the total number of IC

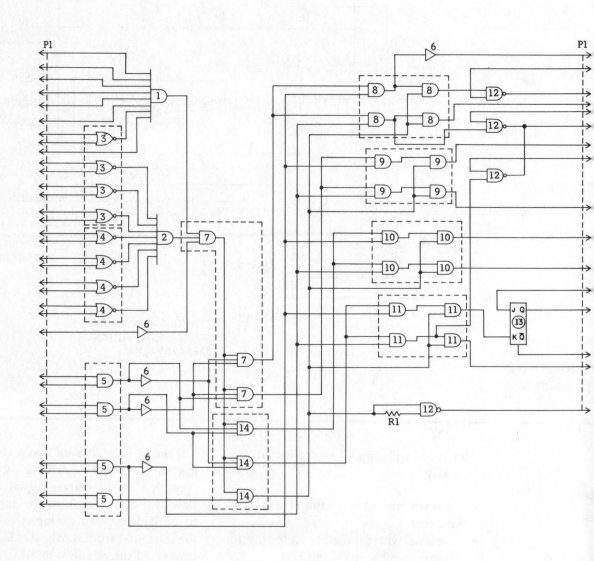

FIGURE 6-3
Logic Diagram

INTEGRATED CIRCUIT GROUPING

packages; and divide the P.C. board square inches by that amount: For example, if there are 12 IC's on a schematic and we had a 4-in. x 6-in. (4 inches wide by 6 inches high) P.C. board divide 12 into 24 = 2 square inches per IC (See Figure 6-4).

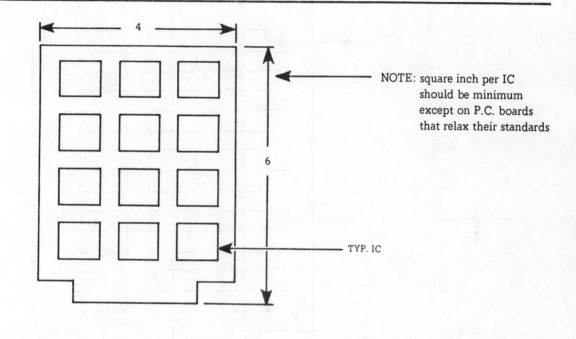

NOTE: square inch per IC should be minimum except on P.C. boards that relax their standards

TYP. IC

FIGURE 6-4 P.C. Board Layout

Now go back to the schematic (Figure 6-3) and try to assign the IC's on the P.C. board.

1. Take a total count of IC's = 14 (using IC spec sheets).

2. Figure total square inches on the P.C. board: 4-in. x 6-in. = 24 square inches.

3. As a rule of thumb, one square inch per IC is minimum for the average P.C. board (use this as a means of figuring the total or maximum number of IC's that can fit on the P.C. board.

4. Count the total number of inputs and outputs to the P.C. board connector to make sure there are enough pins on the connector.

5. Locating the IC's on the P.C. board:

a) A maximum of 24 IC's will fit on the P.C. board.

b) The schematic IC count was 14 IC's total. Divide 14 IC's into 24 square inches = 1.7 square inches per IC (See Figure 6-5).

c) Using Figure 6-5, mark off 5 lines equally spaced horizontally and 3 lines equally spaced vertically. Now at their crossing points, locate 15 IC positions.

Using the schematic and the 15 IC positions, select the positions of the IC's on the schematic.

Things to look for on the schematic regarding the selection of positions on your P.C. board are as follows:

(1) First look at the IC's that go to the connector; they should be placed by the connector end of the P.C. board. In Figure 6-5,

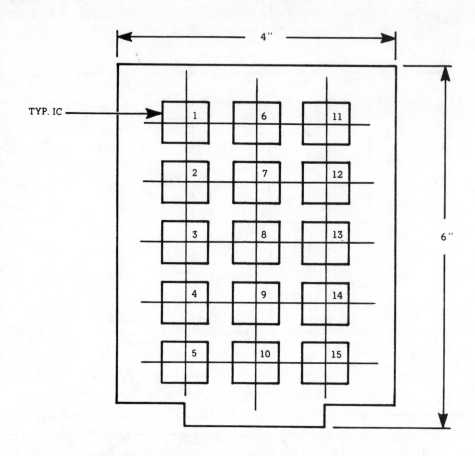

FIGURE 6-5 P.C.
Board Layout

positions 5, 10 and 15 are most preferable; positions 4, 9 and 14 are second in preference.

(2) Since the first row of IC's by the connector positions 5, 10 and 15 are selected, the next IC's to be selected are the ones that connect most to positions 5, 10 and 15. First select the IC that connects to position 5 using position 4, and then the IC that connects to position 4 using position 3, position 2 to 3, and position 1 to position 2. Now back to position 10 for the connections from position 9 to 10 and 8 to 9, etc. If there are not any more connections that connect position 7 to 8

and 6 to 7, then stop on that row and go to the next row, position 14 to 15, etc.

6. At this point, go back to the schematic (Figure 6-3) and use the schematic and the IC count of 14 IC's to assign the IC positions to set this P.C. board up (See Figure 6-6).

a. Looking at the schematic (Figure 6-3), IC3, IC4, IC5 and IC1 are the IC's that connect to the connector the most. Select positions 4, 9 and 14; IC3, IC4 and IC5 are placed in these positions respectively (See Figure 6-6). IC1 should be close to the connector also, so we put IC1 in position 3 because it connects to IC3 and the connector.

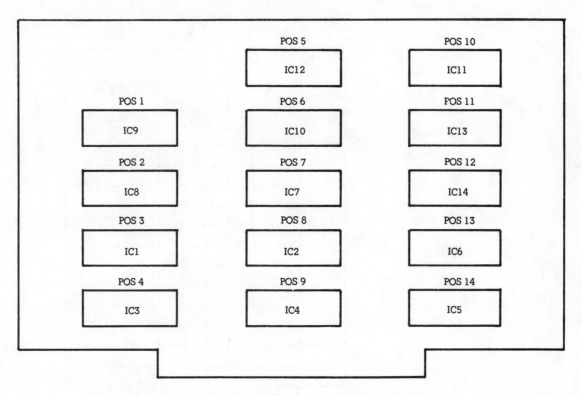

FIGURE 6-6
Final Board
Layout

b. Next, see which IC's are connected to IC4 or position 9. IC2 is connected to IC4 and IC7 is connected to IC2; therefore, position 8 for IC2 and position 7 for IC7 should be selected.

c. Back to the schematic (Figure 6-3) for IC5 or position 14. IC6 connects to IC5, therefore, position 13 should be assigned to IC6. IC6 also goes to IC7 and IC14, which is still close; then IC14 will use position 12.

d. In the schematic the next group of IC's that are common to each other are IC8, IC9, IC10 and IC11, which should be close together using positions 2, 1, 6 and 10.

e. Now the last IC's are out of the common flow. IC12 and IC13 are the last 2 IC's which need positions assigned. First, IC12 connects mostly to IC8, IC11 and IC7, so position 5 would be close to all 3 IC's. Last is IC13, and it should be close to IC11 and would fit very nicely in position 11.

This is just one way to select positions for these IC's; there are several ways and they would probably all be correct. But this will give some idea of how to start and one way to select positions on the P.C. board.

INTEGRATED CIRCUIT GROUPING

101

Chapter

7

DIGITAL LAYOUT

The layout and design of digital P.C. boards deals primarily with IC's, which is unlike the analog layout which deals primarily with discrete components.

Figure 7-1 shows a typical digital P.C. board schematic. The layout should begin by using a .100-in. grid spacing along with IC specifications also shown in Fig. 7-1. The board size would typically be provided by the mechanical engineer. For the examples (Fig. 7-3, 7-7, 7-9), the board size will be 3-in. wide and 4-in. deep. (For further detail on board outlines see Chapter 9). This particular schematic has already had the gates grouped. But, before continuing, take time to review the gate grouping of this schematic.

Step 1 IC #1 - The schematic shows 4 NAND gates. Therefore, there would be no spares.

Step 2 IC #2 - Since some IC's are too complicated to show each gate, they are usually represented by a box on the schematic.

Step 3 IC #3 - There are two 08's which are 2-input AND gates shown on the schematic. The spec sheet shows 4 gates available in the 08, therefore there are two spares.

Step 4 IC #4 - The schematic shows one type 32, 2-input OR gate. The spec sheet shows four available. Therefore there are three spares.

Step 5 IC #5 - The schematic shows two Type 02, 2-input NOR gates. The spec sheet shows there are four available. However, there is one type 27, 3-input NOR gate. The spares in the 02 IC cannot be used for the 27, therefore change the 02 gates to 27, which would make three

3-input positive NOR gates on the schematic, leaving no spares available.

Step 6 IC #6 - The schematic shows five Type 04 inverters. The spec sheet shows six available. Therefore there is one spare inverter.

Step 7 Defining the Board Connector Pin Number. If the engineer does not assign pin numbers to the schematic for the connector P1, then the pin numbers are open for selection. The female connector, typically J-1, is the controlling factor in determining the pin number assignment for P1 on the board. Pin number assignments for this board will be as follows:

All odd numbers 1-25 will be on the component side of the board. All even numbers 2-26 will be on the non-component side.

Step 8 Plotting the IC's on the Board Outline. The 3-in. x 4-in. board outline is 12 square inches which means, 12 IC's could fit on this board, if there is one per square inch. The schematic grouped up to a total of 6 IC's. Therefore, there are two square inches per IC available. It would be logical to arrange these 6 IC's in two configurations. Three IC's across with two deep or two IC's across with three deep as seen in Figure 7-2.

The choice between the two configurations is made by the schematic interconnect. If it is determined to interconnect vertically, place the IC's per Figure 7-2a. If it is easier to interconnect horizontally, use Figure 7-2b.

Step 9 - Use the arrangement of Figure 7-2a and set it up as shown in Figure 7-3. Use Figure 7-2b for plotted locations. Looking at schematic Figure 7-1a, IC 1

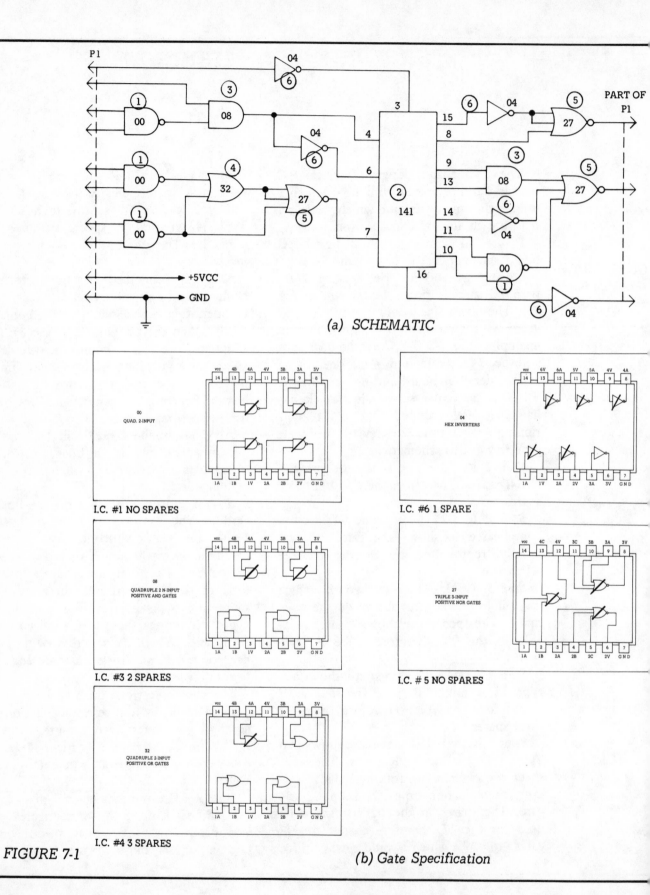

(a) SCHEMATIC

I.C. #1 NO SPARES

I.C. #6 1 SPARE

I.C. #3 2 SPARES

I.C. # 5 NO SPARES

I.C. #4 3 SPARES

FIGURE 7-1

(b) Gate Specification

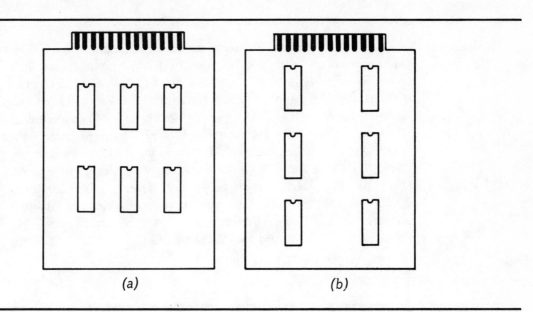

FIGURE 7-2 Two
Possible Board
Layouts

(a)   (b)

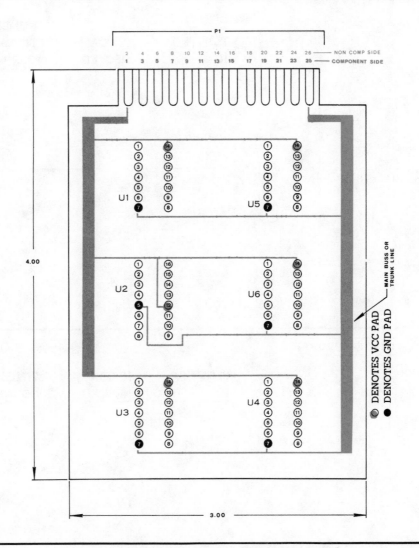

FIGURE 7-3 IC
Layout

and 5 have a lot of connections to P1. Therefore place IC-1 and 5 close to P1. After locating IC 1 and 5, go back to the schematic and try to find which gates interconnect more to IC-5 than any other gates, and which gates interconnect more to IC-1 than any other gates. Establish a chart as in table 7-4 using the IC's that have already been plotted. Starting with IC's 1 and 5.

IC-1 should have IC-2 placed below it because it has more connections on the chart than IC's 3 and 4. Likewise IC-5

should have IC-6 below it because of mor[e] connections. IC's 3 and 4 are left. Choos[e] one of these to be placed under IC-2. Se[e] Table 7-5 below.

It is obvious from the chart that IC-3 ha[s] more interconnections to IC-2, therefo[re] IC-3 should be placed under IC-2. Th[is] leaves IC-4 to be placed under IC-6.

Step 10 Identify the Voltage (Vcc) an[d] Ground Pins on IC-1-6 using the spec shee[t] on Figure 7-1b. IC-1 spec sheet 00 in[d]

PLOTTED IC	IC CON-NECTED	NUMBER OF TIMES CONNECTED
1	3	1
1	4	1
1	2	2
5	6	2
5	3	1
5	4	1
5	2	1

TABLE 7-4
Tabulations of IC Connections

PLOTTED IC	IC CON-NECTED	NUMBER OF TIMES CONNECTED
2	3	3
2	4	0

TABLE 7-5 IC
Connections

DIGITAL LAYOUT

cates that the ground pin is #7 and the VCC is Pin #14. Always use two different colored pencils to distinguish between Vcc and Ground. For example, use red to represent the Vcc pad and black to represent the ground pad. The spec sheets show, that all Vcc pins in this schematic are Pin #14 except IC-2 which is #16 and all ground pins are #7—except IC-2 Pin 5. Now color code them accordingly. Fig. 7-7.

Step 11 - Start putting traces on the printed circuit board. Identify by two different colors which side the traces are on; the component side or the non-component side. Use Pin #1 for Vcc and Pin #25 for ground on P1. Use these pins on this schematic, but they are not necessarily the pins that would be used on other layouts.

Vcc and ground traces are thick traces. The main lines that run down the sides of the board could be even thicker, depending on company standards.

Looking at Figure 7-3, the red pads (Vcc) on IC-1 and 5 are connected with a trace typically min.- of .100-in. thick to the main trunk, which could be thicker. Likewise with row IC-2 and 6 and row 3 and 4. Also in Figure 7-3, black pads (ground) are interconnected on IC's 1 and 5 with a trace that is typically .100-in. minimum to the main trunk, which could be thicker. Likewise with row IC-2 and 6 and row 3 and 4.

Step 12 - The starting point for the interconnection of signal lines (those other than VCC or Ground) is not important. What is important is that the starting point will be the building block from which the rest of the interconnects are based. For example, pick IC-4 which is a type 32, 2-input OR gate (see sheet on Figure 7-6a). Note that on this spec sheet there are four gates to choose, so select one of the four. In the example, select the gate that has Pins 12 and 13 as input pins and 11 as an output pin. Place these numbers on the schematic as shown in Figure 7-6a. This is the beginning of assigning all the pin numbers for

the rest of the gates on the schematic. Decide which three points to connect first. Pins 12?, 13? or 11? In the example 11 is chosen. (It makes little difference which is chosen first). Note that Pin 11 must be connected to a gate called IC-5 (spec sheet - 3-input NOR). The spec sheet shows three gates in this package, select one. Preferably it should be one that is on the same side of IC-4, Pin 11 to simplify the interconnect between IC-5 and IC-4. Also it should be a gate that is close to the bottom of IC-5 because it is closer to IC-4. This is the reason the gate that has input Pins 9, 10, 11 and 8 output is selected. Place these numbers on the schematic. Now Pins 9, 10, and 11 will be interconnected.

Use blue to represent the non-component side of the board which will cross the red color representing the component side of the board. After interconnecting these points on the printed circuit board layout, it is easier to keep track of which points are interconnected on the schematic by using a yellow pencil to cross out the lines as they are interconnected. This ensures no missing interconnects.

Now start interconnecting to another point, preferably a gate that is numbered. For example, IC-4 has Pins 12 or 13 and IC-5 has 8 or 11. Pick IC-5, Pin 11, which is also connected to two other points. One of these two has a number which was assigned, IC-4, Pin 13.

There is now one unknown point to interconnect on IC-1. First, interconnect IC-5 Pin 11 to IC-4 Pin 13. Using the underside of the IC's. This way, is not necessary to overlap the blue lines which were just laid down. Now select a gate on IC-1 to interconnect to the blue line out of IC-5, Pin 11. The closest gate to this line in IC-1 (2-input NAND) would be Pins 9, 10 input, 8 output. Assign this to the schematic and using a feed-through in the second aisle (of Figure 7-7) place it under Pin 8, IC-1. Cross the red line out of Pin 8 with a blue line to the feed-through pad. Next using red,

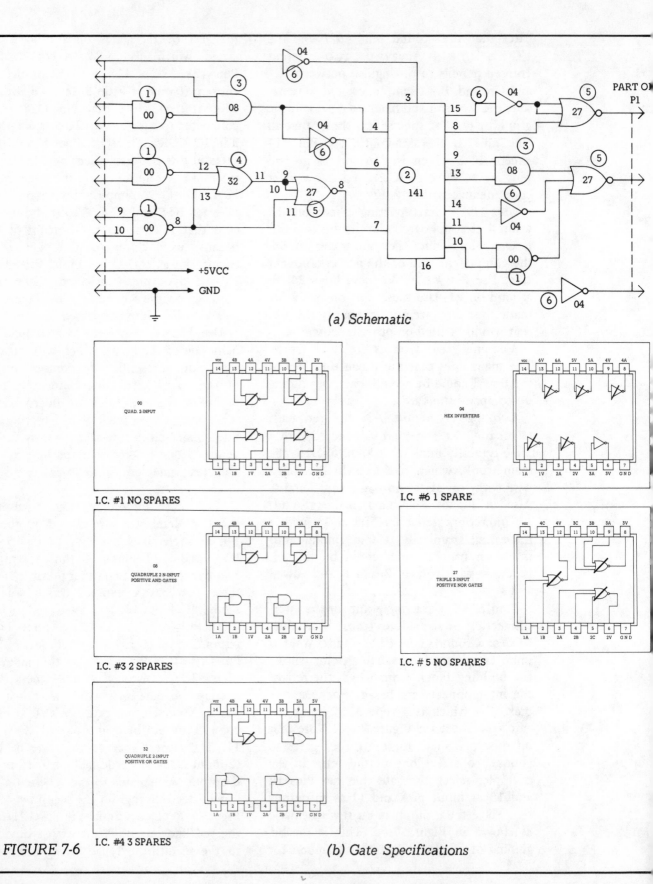

(a) Schematic

I.C. #1 NO SPARES

00
QUAD. 2-INPUT

I.C. #6 1 SPARE

04
HEX INVERTERS

I.C. #3 2 SPARES

08
QUADRUPLE 2 N-INPUT
POSITIVE AND GATES

I.C. # 5 NO SPARES

27
TRIPLE 3-INPUT
POSITIVE NOR GATES

I.C. #4 3 SPARES

32
QUADRUPLE 2-INPUT
POSITIVE OR GATES

(b) Gate Specifications

FIGURE 7-6

DIGITAL LAYOUT

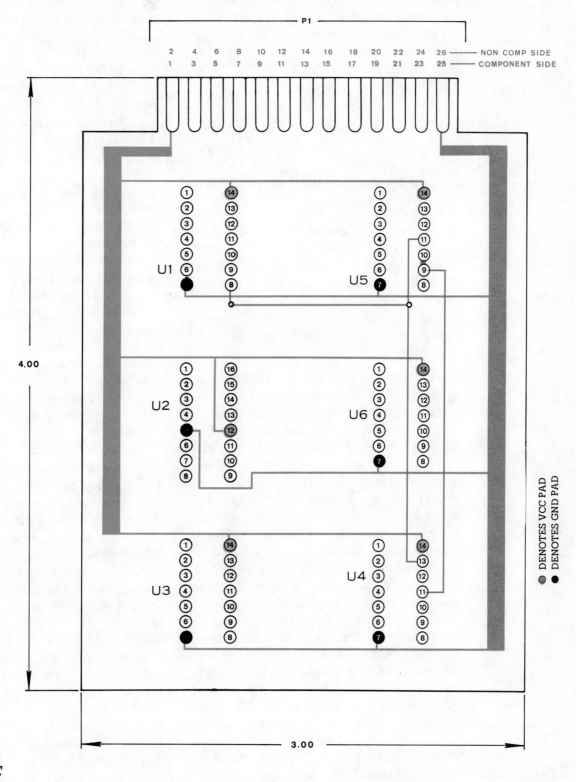

FIGURE 7-7 IC
Layout

draw a line over to the blue line out of IC-5, Pin 11. Now place a feed-through pad connecting IC-5, Pin 11 to IC-1, Pin 8. With a yellow pencil cross out the lines on the schematic interconnecting IC-5, Pin 11, IC-4, Pin 13 and IC-1, Pin 8.

Now that we have given you some insight to the interconnecting phase, such as:

- Establishing board size and connector size.
- Grouping gates.
- Assigning Pin numbers to gate inputs and outputs.
- Establishing signal line colors for component side and non-component side or the printed circuit board.
- Establishing direction of signal lines (different color each side) for the component and non-component side of the board.
- Establishing connector pin number assignment.

Now proceed with the rest of the interconnects using schematic Figure 7-8a with a brief explanation of each interconnect to be shown on Figure 7-9. Connect IC-4, Pin 12, which is the only unconnected pin left on this gate, to an output pin on IC-1 (spec sheet 00 2-input NAND gate). Find the shortest distance between these two points, keeping in mind that an output pin must be chosen. Examination of sheet 00 shows that the choice is limited to Pins 3, 6 and 11 because 8 had been previously used; Pin 11 is selected as shown in Figure 7-9. At this point assign the selected pin numbers to the schematic. Assume from this point on that each time a gate pin number is interconnected, it will be assigned to the schematic. For the following interconnects see Figure 7-9. For the next interconnect put IC-5, Pin 8 to IC-2, Pin 7. Next interconnect IC-2, Pin 6 to an output on IC-6, (spec sheet 04, inverter). To obtain the shortest connection, pick 6 or 8; Pin 8 is selected, making the input side Pin 9. Next interconnect IC-2, Pin 4 to IC-6, Pin 9

and to an output on IC-3 (spec sheet 08 Out of IC-2, Pin 4, the shortest inte connect to IC-3 would be Pin 3. Once Pin on IC-3 is picked it automatically assig Pin 1 and 2 of the same gate.

Next interconnect, IC-3, Pin 2 to IC output pin. Available pins to choose fro are Pins 3 and 6. Choose Pin 6 since it the closest available pin which automa cally assigns Pins 4 and 5 to the input si of the same gate.

Next interconnect IC-2, Pin 3 to an ou put pin or IC-6. Available pins to choo from are Pins 2, 4, 8, 10 and 12. Choo Pin 2 which automatically assigns Pin 1 the input side.

Next interconnect IC-2, Pin 15 to IC input side. Available pins to choose fro are 3, 9, 11 and 13. Choose pin 3 whic automatically assigns Pin 4 for the outp pin.

Next interconnect IC-2, Pin 8 to IC input side. Closest available input pins o spec sheet 27 are pins 3, 4 and 5. Choo pin 5 which automatically assigns Pin 3 an 4 to the remaining inputs and Pin 6 to th output.

Next interconnect IC-5, Pins 3 and connect together and then connect to IC-Pin 4. Then interconnect IC-2, Pin 9 t IC-3 spec sheet 08 to input and, input sid Available pins to choose from are 4, 5, 10 or 12 and 13. Choose Pin 13 whic automatically assigns Pin 12 to the re maining input and 11 to the output. The interconnect IC-3 pin 12 to IC-2, Pin 13.

Next interconnect IC-3, Pin 11 to a input on IC-5 (spec sheet 27 3-inpu NOR). Available pins to choose from are 2 or 13. Choose Pin 2 which automaticall assigns Pins 1 and 13 to the remainin inputs of the same gate and Pin 12 to th output. Next interconnect IC-5, Pin 1 t IC-6 output side; available pins to choos from are 6, 10 or 12. Choosing Pin 6 aut matically assigns Pin 5 to the output side the same gate. Next interconnect IC-6, Pi 5 to IC-2, Pin 14.

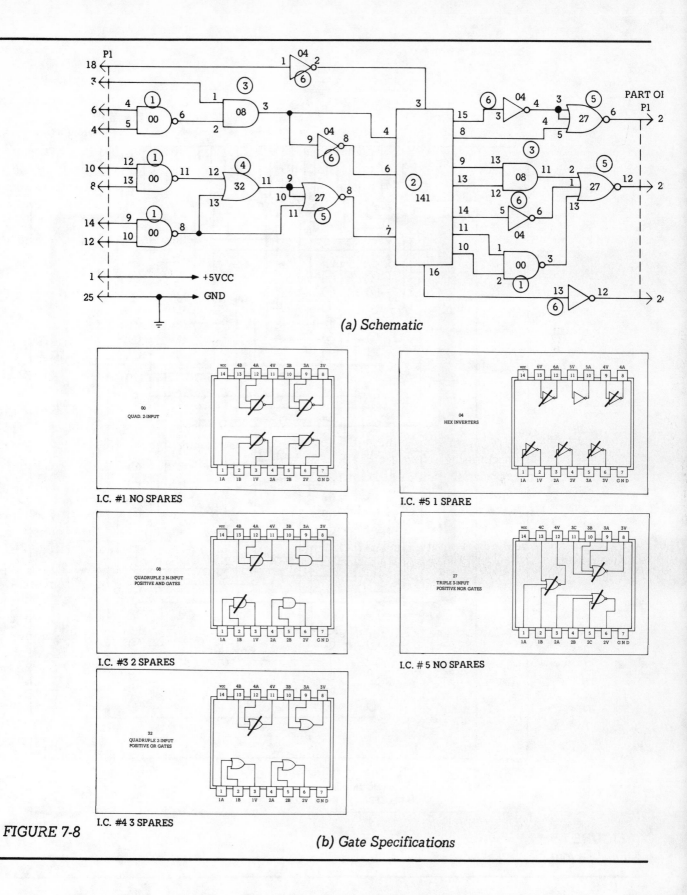

(a) Schematic

I.C. #1 NO SPARES

I.C. #3 2 SPARES

I.C. #4 3 SPARES

I.C. #5 1 SPARE

I.C. # 5 NO SPARES

FIGURE 7-8

(b) Gate Specifications

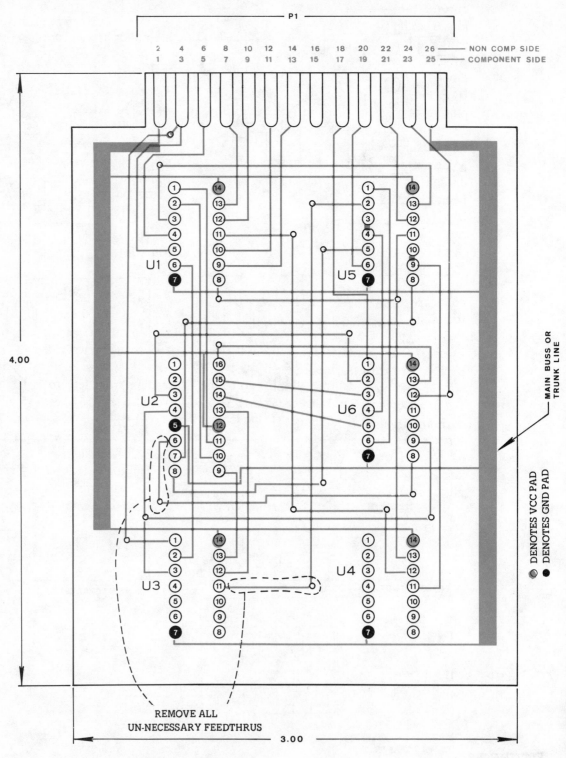

FIGURE 7-9
Final Layout

Next interconnect IC-5, Pin 13 to an output on IC-1. Pin 3 is the only remaining output pin. Which automatically assigns pins 1 and 2 to the input side of the same gate.

Next, interconnect IC-1, Pin 1 to IC-2, Pin 11; IC-1, Pin 2 to IC-2, Pin 10; and IC-2, Pin 16 to an input on IC-6. Available pins to choose from are 11 and 13. Choose 13 which automatically assigns 12 to the output side of the same gate.

The remaining interconnects to be made on the schematic all terminate to the connector P1. In the schematic, P1 does not have preassigned pins for these connections. Therefore as routed the traces from the IC gates to the connector P1, to choose the closest available pin on the odd numbered side or the even numbered side of the connector. Once a pin is chosen, assign it to the schematic.

See Figure 7-8a for the schematic and Figure 7-9 for the interconnections to connector, P1. On IC-1 we have Pins 4, 5, 12 and 13 and 9 and 10 that all must go to the P1. Starting with Pin 5 on IC-1, route 1 - up to Pin 4 on P1. Next connect IC-1, Pin 4 to Pin 6 on P1; IC-1, Pin 13 to Pin 8; IC-1, Pin 12 to Pin 10; IC-1 Pin 10 to Pin 12 of P1; and IC-1, Pin 9 to P1, Pin 14. This connects all the inputs on IC-1 to P1.

Then, IC-3, Pin 1 must be connected to a pin on P1. The closest available pin is Pin 3 on P1. Note on Figure 7-9 that the blue trace we used to route IC-3 Pins to P1 must have a feed-through close to P1, pin 3 to connect to the red side which is the odd numbered side of the connector.

The next connection is IC-6, Pin 1 to P1. The easiest pin to connect to is Pin 18. Following this, IC-5, Pin 6 to P1 is connected to the closest available pin on P1, Pin 20. The next connection IC-5, Pin 12 to P1. The closest available connection to P1 is Pin 22. At this point all the interconnects on the schematic are completed. One of the last steps remaining on the layout is to remove all unnecessary feed-throughs. For examples, see Figure 7-9.

Also, the IC numbers which were assigned in the original grouping of the schematic could be changed to put them in numerical order on the printed circuit board layout. For example, Figure 7-9 has IC-1 in the upper left hand corner with IC-5 to its right. These could be rearranged to have IC-2 in place of 5. In other words, reading from left to right, top to bottom would be IC-1, 2, 3, 4, 5 and 6 or 1 where 1 is, 2 where 5 is, 3 where 2 is, 4 where 6 is, 5 where 3 is, and 6 where 4 is. If the IC numbers are changed as listed, match new IC numbers to the schematic. In other words, when the gates for IC-5 were called out on the schematic, change the 5 to 2 and etc.

Chapter

8

MULTILAYER P.C. BOARDS

MULTI-LAYER PC BOARDS

This chapter introduces the design of multilayer PC boards that accommodate up to 2½ times the component density of single layer and double sided boards, and many, many times their wiring density.

Multilayer boards are used as fixed-position "backplanes" through which component-laden PC boards are interconnected. They are also used as plug-in "mother boards" which hold components and edge-connected "daughter boards," and are also used individually, to facilitate highly complex circuits.

Layer-to-layer registration is critical, and design difficulty is not to be underestimated. This chapter defines the tradeoffs required for multi-layer boards and details the step-by-step procedure for designing them.

MULTI-LAYER P.C.B. CONSTRUCTION

Multilayer P.C. boards contain three (3) to as many as twenty (20) or more layers, each composed of a single plane of copper. The copper laminated to one side of the substrate is referred to as the layer. Each layer is etched by the board manufacturer from the reduced artwork. These etched layers are bonded together using materials of glass fiber and epoxy resin. These bonding materials are referred to as "prepeg" or "B"-stage, Figure 8-1.

The registration of these many layers is very critical. Therefore, the responsibility of the P.C. designer is to use great care in the preparation of the artwork; this point cannot be over stressed. The accuracy in locating these many layers in relationship to each other is determined by the width of the annular rings which are used to interconnect the various points in each layer. Misalignment of layers could result in either short circuits or open connects within the multilayer board. Interconnections of the many layers are made possible through the use of plated through holes, (Figure 8-2) as done with the conventional double-sided printed wiring board.

The plating method for making interconnections through holes in the dielectric substrate of P.C. boards has been established and proved out in double-sided P.C. boards. The use of this technique on P.C. boards of more than two (2) layers is an extension of the state-of-the-art. This appears to be the most widely accepted technique, and in practice it must account for 80-90 percent of the multilayer P.C. boards used today. Figure 8-3A and 3-B shows a typical cross sectional view of a plated through-hole in a multilayer P.C. board connecting the inner layers to the feed-through by means of the plated wall on the hole.

The following are some advantages and disadvantages to using a plated-through-hole in a multilayer P.C. board:

A. *Advantages*

The principal advantages of the plated-through-hole method is the increase in miniaturization which is not

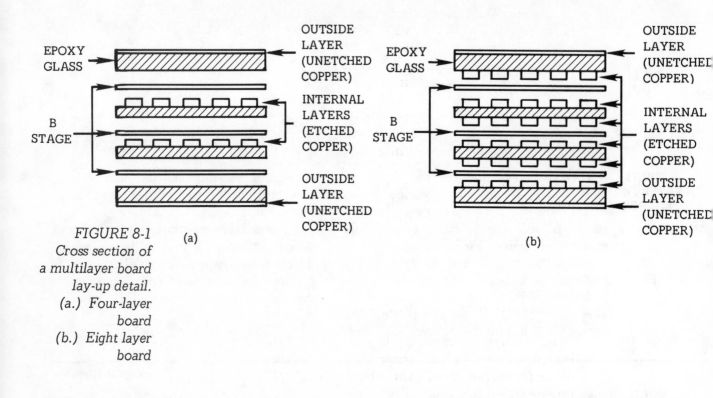

EPOXY GLASS

OUTSIDE LAYER (UNETCHED COPPER)

B STAGE

INTERNAL LAYERS (ETCHED COPPER)

OUTSIDE LAYER (UNETCHED COPPER)

EPOXY GLASS

OUTSIDE LAYER (UNETCHED COPPER)

B STAGE

INTERNAL LAYERS (ETCHED COPPER)

OUTSIDE LAYER (UNETCHED COPPER)

(a)

(b)

FIGURE 8-1
Cross section of
a multilayer board
lay-up detail.
(a.) Four-layer
board
(b.) Eight layer
board

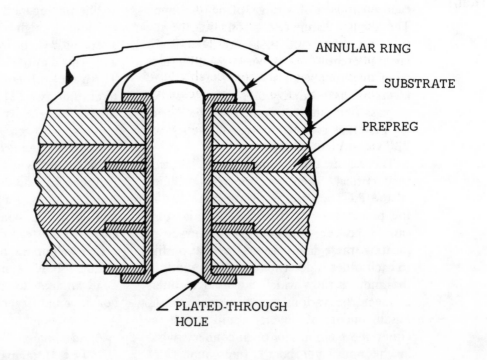

ANNULAR RING

SUBSTRATE

PREPREG

PLATED-THROUGH HOLE

FIGURE 8-2
Plated-through-
hole in a multi-
layer board.

MULTILAYER P.C. BOARDS

obtainable by other methods. The layout is made easier by utilizing standard grid centers. The reasons for using the standard grid centers are as follows: registering the artwork from layer to layer, ease of taping the art masters, and for realigning the layers in the manufacturing process.

The plated-through interconnection technique has probably enjoyed the greatest acceptance to date, since most P.C. boards benefit from its compactness and its ability to take from two (2) to twenty (20) layers of interconnections without affecting the space requirement adversely.

B. *Disadvantages*
The designer must carefully evaluate potential manufacturers who demonstrate background capability in making plated-through-hole connections on a production basis. The plated-through-hole technique is a fine-art of printed circuit manufacturing, and requires considerable technological experience and quality control. If there are disadvantages to the plated-through-hole technique, they might be listed as a considerably lower production yield than by techniques utilizing mechanical means, slower production, repairability and reuseability.

The most common use of the plated-through-hole is seen in Figure 8-3C. The use of this technique is important because of the frequent need to connect only the outer layer of the multilayer P.C. board. This is accomplished by providing a clearance around the hole on the inner layers. This is typical when a signal passes through a ground or voltage plane. In manufacturing the P.C. board, a concentric circle of copper is etched away leaving a sufficient diameter around the hole. This

minimum clearance will assure that any misregistration during fabrication will not result in shorting of traces or voltage and ground planes.

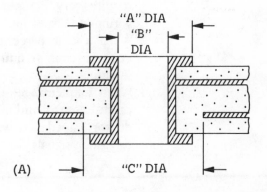

(A)

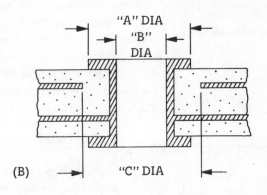

(B)

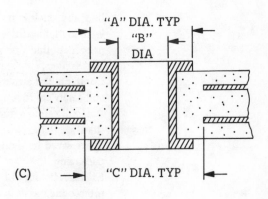

(C)

FIGURE 8-3
(A & B) show
typical connection
of inner layers to
feed-through.
(C) shows the
clearance around
the feed-through.

NEED FOR MULTILAYER P.C.B. Many factors will influence the need for a multilayer printed wiring design. Some of these are as follows:

1. When the interconnections in sub-assemblies are very complex and will require very complicated and costly wiring or harnessing.
2. When a large percentage of the interconnections require coupling or shielding.
3. When frequency requirements call for careful control and uniformity of conductor impedance with minimal distortion in signals.
4. When interconnections, size, and weight are critical, as in military, airborne, missile, and space applications.
5. When the quantity of components discrete, integrated circuits, and terminals exceed the normal design area and close spacing is a necessity.
6. When the need to connect a large number of interconnections from one pc board to another, as in back planes of complex systems.

ADVANTAGES AND DIS-ADVANTAGES When evaluating whether to use multilayer boards, a comparison should be made among the different forms of printed wiring that could be used in the end product. A determination should be made as to how many double-sided boards versus multilayer boards could fit into the allotted design area. Other factors are cost of manufacturing, assembly, and maintenence time of the end product.

The following is a list of some possible advantages and disadvantages when deciding to use multilayer boards.

ADVANTAGES	DISADVANTAGES
1. Wiring errors eliminated	1. Special tools and skills required for board repairs
2. Uniform electrical characteristics (impedence, coupling, and shielding).	2. High cost in small quantities
3. With high quantities and proper tooling a possible cost savings.	3. Design time is longer.
4. Combining of structural and electrical functions.	4. Lead time for board fabrication is longer.
5. Assembly time reduced (hard wiring).	5. Board warpage (thermal sensitivity).
6. Space saved by tighter packaging.	6. Inspection of boards is difficult and generally requires microsectioning.
7. Higher reliability of interconnections.	7. Very costly and difficult to change completed boards.
	8. Open connections from *SMEAR* in plated-through-hole.

PACKING DENSITY The use of multilayer boards is particularly important when the maximum component densities possible with double-sided boards must be exceeded. For instance, typical IC densities on double-sided boards are on the average of one IC per square inch, whereas this is 2 IC's per square inch on multilayer board. This is also possible when using other components, where the double-sided board generally uses four (4) TO case sizes per square inch, this can be increased to over eight

MULTILAYER P.C. BOARDS

(8) TO case sizes per square inch when multilayer designs are used. These densities in component packaging can be increased to twice the amounts mentioned when the components are mounted so that the bodies touch each other in final assembly. This method does often require more layers for the interconnecting circuitry.

Many standards and specifications that apply to single and double-sided boards also apply to multilayer P.C. boards. Although there are several which apply primarily to multilayer P.C. boards. A few of the most commonly used are:

A. Institute of Printed Circuits Specifications (IPC)
 1. IPC-ML-910: Design and end product specification for Rigid Multilayer boards.
 2. IPC-ML-950: Performance specifications for Multilayer Printed wiring boards.
 3. IPC-ML-975: End product documentation specification for Multilayer Printed wiring boards.

B. Department of Defense Specifications
 1. MIL-STD-1495: Multilayer Printed wiring boards for electronic equipmennt.
 2. MIL-P-55617: Plastic sheet, thin laminate, metal clad, for printed wiring, primarily for multilayer.
 3. MIL-G-55636: Glass cloth, resin preimpregnated, B-stage, for multilayer printed wiring boards.
 4. MIL-P-55640: Printed wiring boards, multilayer, plated-through holes.

When considering the use of multilayer P.C. boards over the conventional single- or double-sided board, a great deal of pre-design time is required. It should be remembered that there is an optimal limit to the size and cost of any discrete unit or functional subassembly/integrated circuit. It becomes uneconomical and inefficient to increase the size or complexity beyond that limit, because the unit then becomes too costly, specialized or not easily interchangeable. Therefore, a great deal of packaging requirements must be considered. If the investigation into the various approaches indicates a need for a multilayer P.C. board to achieve the desired component density, then the following list of design factors should be evaluated for the multilayer board:

A. Design Considerations:
 1. Allotted board area
 2. Number of layers required
 3. Number of hole sizes
 4. Total board thickness
   a. Inner layer thickness
 5. Width of conductors (internal & external)
 6. Terminal areas (internal & external)
 7. Spacing considerations
   a. Center-to-center component mounting holes
   b. Terminal area to terminal area
   c. Terminal area to conductor
   d. Conductor to conductor
   e. Terminal area to board edge
   f. Conductor to mounting holes
   g. Conductor to board edge
 8. Tolerances (design)
 9. Drawing requirements
 10. Special components
 11. Special design requirements

MULTILAYER P.C. BOARDS

12. Terms and definitions (unique to design)
13. Plating (type and thickness)

B. Performance & environmental considerations
   1. Tensile strength
   2. Flexure strength
   3. Vibration
   4. Mechanical shock
   5. Temperature cycling
   6. Thermal shock
   7. Moisture resistance
   8. Fungus resistance
   9. Corrosion resistance (salt spray)
   10. Twist or warp
   11. Solderability & resolderabilit
   12. Insulation resistance
   13. Dielectric strength (voltage)
   14. Temperature storage (high-low)
   15. Conductor temperature rise
   16. Conductor resistance
   17. Flame resistance
   18. Altitude requirement (high-low)
   19. Machineability

BOARD DESIGN LAYOUT

When the complexity of a schematic diagram indicates the need for a multilayer board and the many design considerations have been weighed, one of the first problems to solve is the distribution of signals, ground, and voltage planes between the number of layers required. There are many possible solutions to this problem, therefore it becomes a trial-and-error method on the part of the drafter or designer to achieve the design with the minimum number of layers. In some companies, computers are being utilized to help determine the simplest multilayer design which helps eliminate some of the trial-and-error method.

Before the actual design layout can be started, ground rules about conductor sizes must be developed. Preparation of the artwork can be accelerated if a uniform conductor width is used throughout the entire board design, although, some narrowing down in critical areas is permitted occasionally. Also, if power or ground distribution buses are used instead of planes, they should be made as wide as possible to minimize voltage drop and maximize shielding.

Once the board size has been determined, all module positions distributed, pad or hole location fixed and input-output methods selected, the formal layout of the board can be started. The layout should contain all the above information as well as mounting holes to secure the board to the frame or holes for mounting other connectors or components to the board. It also important to indicate any unusable areas on the board where no conductor should be placed. Figure 8-4 shows some typical examples of unusable area. This will prevent the possible danger of the conductors on the inner layers from being cut through or damaged during hole drilling, a final operation prior to assembly. Spacing of conductors is of major importance. It is best to determine the maximum number of conductors that can be spaced properly between two terminal areas in various areas of the board, and then stick to rules. Figure 8-5 gives an example of maximum number of traces that might run between the pad of an IC.

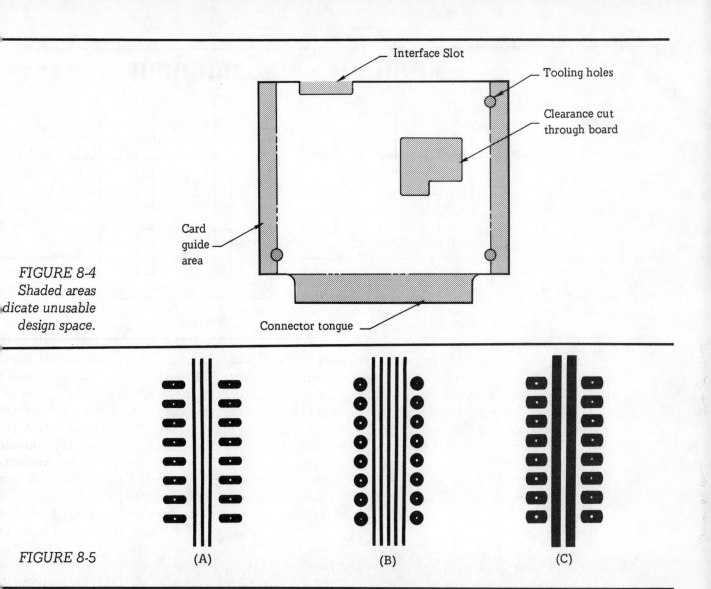

FIGURE 8-4
Shaded areas
indicate unusable
design space.

Interface Slot

Tooling holes

Clearance cut
through board

Card
guide
area

Connector tongue

FIGURE 8-5   (A)   (B)   (C)

It should be noted here that one person should work on the design from start to finish, and the work should not be divided among several designers or drafters. The checking should be done by someone other than the designer.

At this point the layouts of both a double-sided P.C. board and a multilayer P.C. board will be compared. It will be assumed that the necessary grouping of the logic gates and pins for gates has been assigned, color codes for the different voltages, grounds and layers selected, and the number of layers determined. The steps used to achieve these designs are similar to those discussed in Chapter 7. Although these layouts appear the same. there are some very distinct differences in the final layouts.

When comparing the multilayer P.C. layout to the double-sided P.C. layout (illustrated in Chapter 7) the first thing that appears different is that the ground and voltage IC pads on the multilayer design are color-coded and not laid in as bus lines as illustrated in the double-sided design. Figures 8-6 and 8-7 show this difference. The reason for this is that the voltages and ground will be on the inner layers of the multilayer P.C. board. The connections of these voltages and ground are made possible through feed-throughs which pass through the inner layer of the P.C. board.

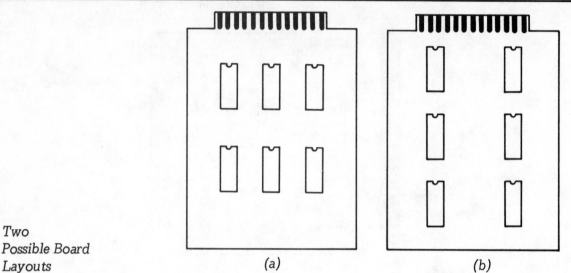

Two
Possible Board
Layouts

(a)   (b)

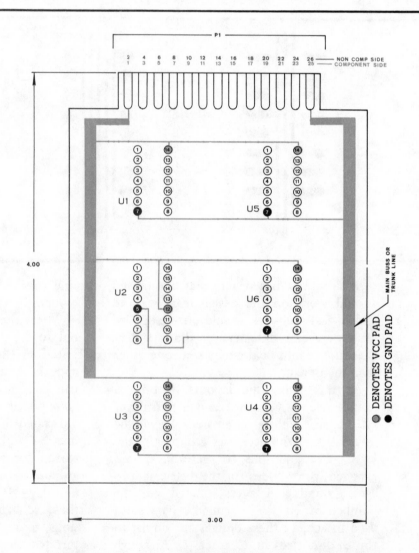

FIGURE 8-6
Shows the ground
and voltage traces
layed into the
design.

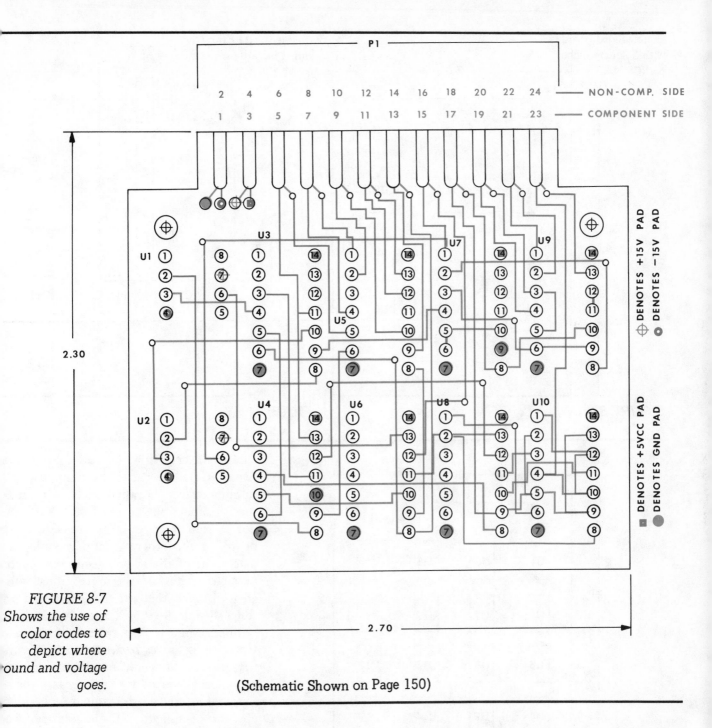

FIGURE 8-7
Shows the use of color codes to depict where ound and voltage goes.

(Schematic Shown on Page 150)

The technique for placing these voltages and grounds on the inner layers provides more space for ICs and signal traces. This becomes very apparent when the areas of the two boards are compared. See Figure 8-8 and 8-9 for board dimensions. The amount of area allotted for the double-sided board is one and two square inches per IC, whereas the area alloted for the multi-layer board is about .6 square incher per IC. This is an increase of more than two and a half times in the multilayer design.

With the ICs placed close together as seen in Figure 8-9, the next thing to notice is that the aisle-ways are much closer together. Therefore, if the techniques used in the double-sided boards were used, this multilayer design would not be possible.

MULTILAYER P.C. BOARDS 127

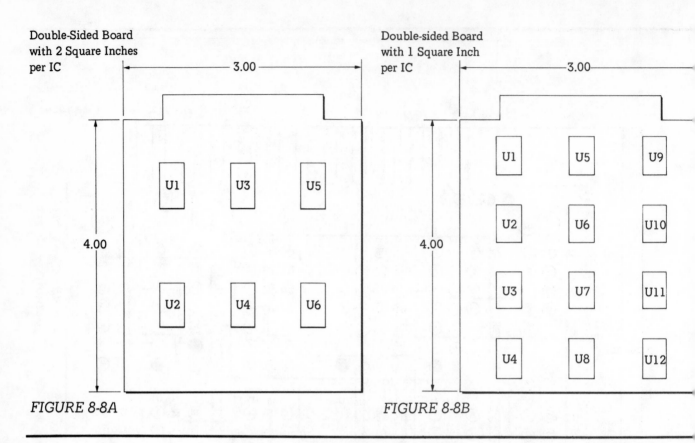

Double-Sided Board
with 2 Square Inches
per IC

3.00

4.00

U1 U3 U5

U2 U4 U6

FIGURE 8-8A

Double-sided Board
with 1 Square Inch
per IC

3.00

4.00

U1 U5 U9

U2 U6 U10

U3 U7 U11

U4 U8 U12

FIGURE 8-8B

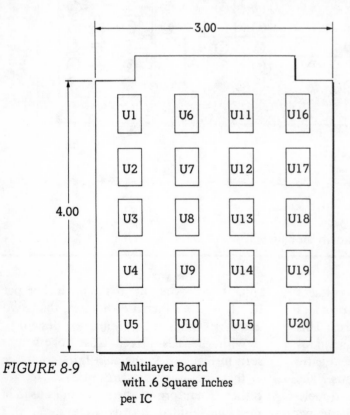

3.00

4.00

U1 U6 U11 U16

U2 U7 U12 U17

U3 U8 U13 U18

U4 U9 U14 U19

U5 U10 U15 U20

FIGURE 8-9

Multilayer Board
with .6 Square Inches
per IC

Hence, the technique of running traces be
tween IC pads is used. The main idea whe
running traces between pads is to mak
these connections as short as possible an
not to run a trace from one side of th
board to the other. Careful grouping an
placement of the IC will insure the shorte
possible routes. If the traces must run
great distance, then use the aisle-way if po
sible. Figure 8-10 will show examples o
both techniques. A very important reaso
for running traces between IC pads is tha
the amount of feed-throughs is eliminate
or at least reduced which increases the de
sign real estate. At this time it is also advi
able to eliminate all unnecessary fee
throughs. Figure 8-11 shows some possibl
feed-throughs that can be removed.

Figure 8-12 shows three (3) tooling hole
that have been placed within the boar
outline. These tooling holes could be use
when the different layers of the P.C. boar
are laminated together during final assembly

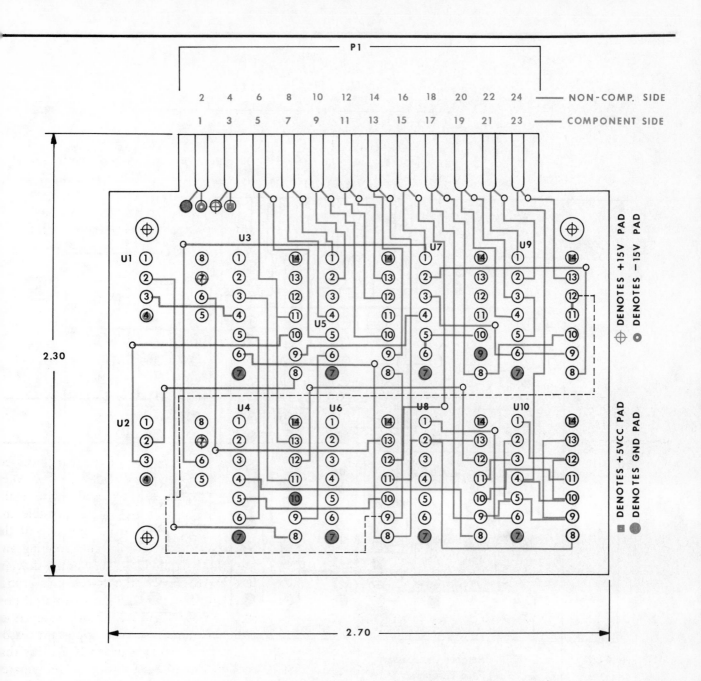

FIGURE 8-10

(Schematic Shown on Page 150)

These tooling holes are important and should not be overlooked when making the layout. It is a recommended design practice to locate at least three (3) holes in such a manner that the board cannot be misaligned during lamination and assembly. Also provide enough space around these tool-ing holes so that during drilling no traces will be cut or destroyed.

Once the penciled layout of the design is completed and all design parameters are met, the next step is to prepare the art-work.

MULTILAYER P.C. BOARDS   129

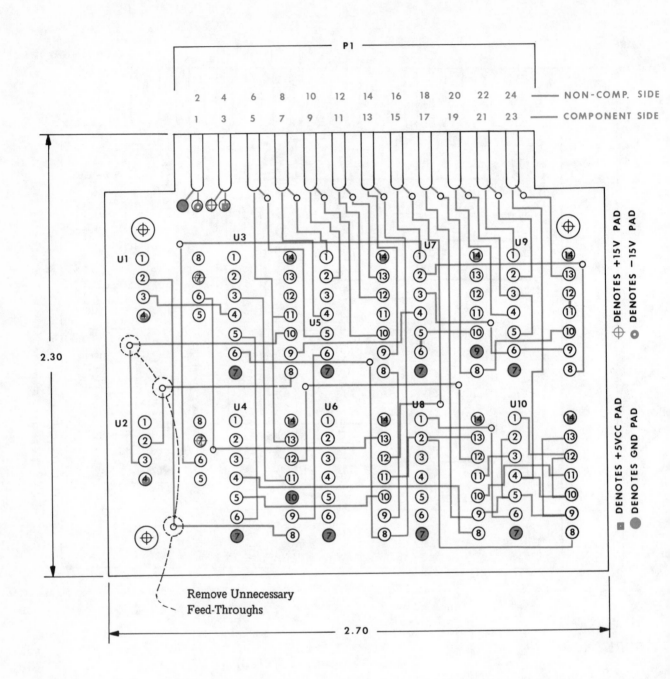

FIGURE 8-11

(Schematic Shown on Page 150)

MULTILAYER P.C. BOARDS

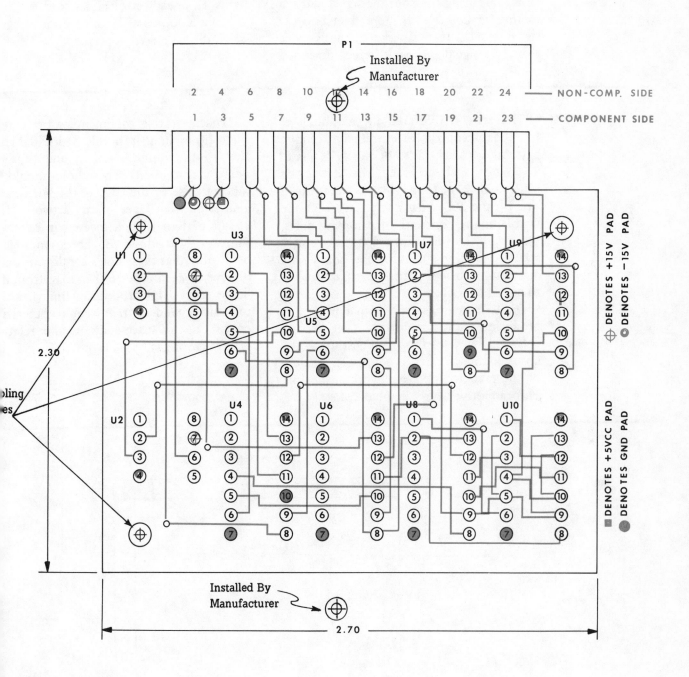

FIGURE 8-12
shows the possible
locations of tooling
holes in a typical
board design.

(Schematic Shown on Page 150)

DESIGNER FOOTNOTE The preceding discussion has described the procedures for laying out a four (4) layer P.C.B. What must also be considered is how to lay out using color for the signal paths, a multilayer board with more than two (2) signal planes. If the design requires more than two signal planes, it is recommended that a separate sheet of mylar be used f each additional signal plane. This will eli nate the confusion of having all the sig paths on one sheet of mylar, and will p vide a less confusing layout for the design when the art master is prepared.

ARTWORK The artwork required for a multilayer P.C. board consists of extremely accurate patterns for each layer of the P.C. board. To obtain such accuracy, the artwork is made two (2) or four (4) times the finished size of the P.C. board. The layers of the board require registration or tooling holes, the board outline dimensions, extremely accurate outline marks of the board, and dimensions to which the board artwork is to be reduced.

One of the most critical parameters in producing multilayer boards is registration between layers. Proper registration of layers in the finished product depends upon the precision of the artwork or tape-up. This cannot be overly emphasized.

To achieve this, tooling holes, preferab three (3) in number, should be located re tive to the outer edges of the finish board (Figure 8-13). These holes should located close to the edge of the board, a appear on each layer of the artwork. T circuit pattern is then located in relatio ship to these three (3) holes. In Figu 8-13, note that these holes are dimensione relative to each other rather than from t board edges. This procedure eliminates t tolerance build-up that would otherwi contribute to inaccuracy in the relati placement of the board layers.

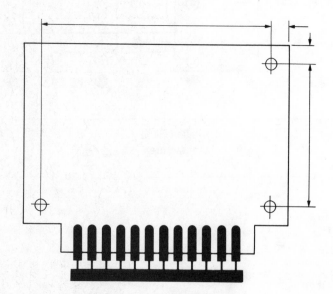

FIGURE 8-13

In addition to the three (3) tooling holes, each layer of the artwork should also include two (2) reduction targets with the exact reduced dimension and tolerance between them. This serves as a guide for the photographer when reducing the artwork (Figure 8-14). In some cases two sets of reduction targets and dimensions are used on the artwork (Figure 8-15). The reason for the two sets of targets and dimensions is that as the size of the board in one direction becomes greater than 6.00 @ 1:1, it becomes necessary for the photographer to check in each direction of measurement for errors in camera parallax.

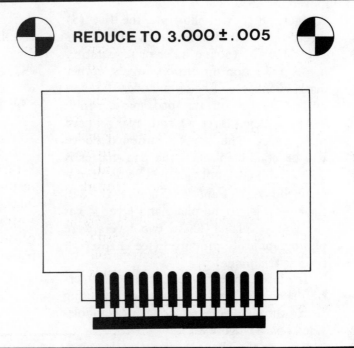

REDUCE TO 3.000 ± .005

FIGURE 8-14

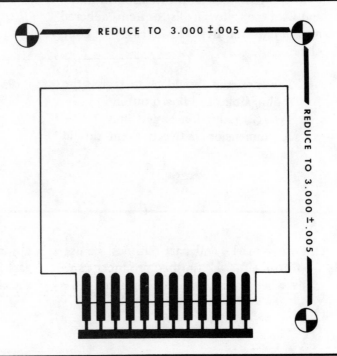

REDUCE TO 3.000 ±.005

REDUCE TO 3.000 ±.005

FIGURE 8-15

One of the most critical responsibilities of the drafter or P.C. designer is the artwork preparation itself. The drafter or designer must make sure that each sheet has registration marks or tooling holes, reduction targets and dimensions, and that each layer is identified correctly. These items are very important and must not be overlooked.

Figure 8-16 will illustrate the five (5) sheets of a multilayer artwork. Note: The three (3) tooling holes or registration marks, reduction dimension targets, corner marks, connector fingers and over all board outline appear on the pad master only. Note the taped layer #1 and layer #4 have only traces. The items mentioned above will be combined when the pad master is overlayed and photographed. See photographed layer # 1 and # 4 Figure 8-16A&B. Layer # 2 is ground plane and layer # 3 is the voltage plane. These two layers were photographically produced (see chapter 12 for this technique).

The interconnection of the various layers of the multilayer board is made by using plated-through holes as done in a double-sided board (Figure 8-2).

Keeping the above information in mind, and with the following list of items adhered to, a multilayer artwork will easily be completed.

1. The first step should be to layout the tooling holes and board outline.
2. Two (2) reduction targets and the exact dimension between them should be included.

3. Layout the main component pads (o hole) locations for IC's patterns, resis tors, capacitors, etc.
4. Photograph (1 x 1 film positive) o contact print the 2:1 or 4:1 tape-u master. Now make one (1) copy fo each layer of artwork.
5. Layout the conductor patterns for th internal layers first, eliminating an unwanted feed-through pads.
6. Layout the conductor patterns for th two (2) external layers last.
7. Check registration of all layers an clearance for all conductor pattern
8. The following considerations must a so be given careful attention to insur a satisfying end product:
   a. Conductor layout—tape of prop widths should be selected to pr duce the desired conductor widt after reduction.
   b. Pad and hole layout—pads are usua ly round, although square and sp cially shaped pads are sometim used (see hole formula, page 59, fo selecting proper pad size.) Whe ground and voltage planes appear inner layers, the clearance hole r quired to eliminate unwanted co nections to the plated-through hol must be clearly marked. This is e sential as these are fabricated in t plane prior to laminating.
9. Identify each layer of artwork starti with sheet one (1) component sid sheet two (2) voltage, etc. This iden fication is to be placed outside t board outline.

Corner marks and part outlines are useful features for the finished products, especially when hand-tooling does not exist. As a minimum, corner marks should appear each layer of the artwork; these simpli the checking processes.

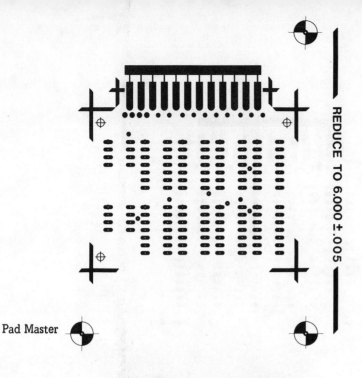

Pad Master

REDUCE TO 6.000 ±.005

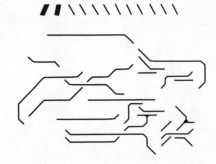

Layer # 1
Component Side
Traces

Layer # 2
Ground Connections

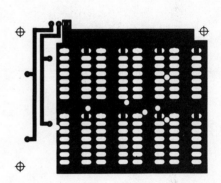

Layer # 3
Voltage Connections

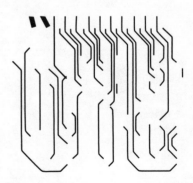

Layer # 4
Non-Component
Side Traces

FIGURE 8-16

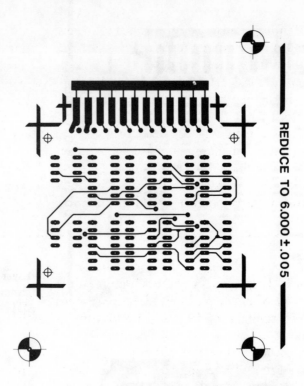

FIGURE 8-16A
Component side
traces & pad
master.

REDUCE TO 6.000 ±.005

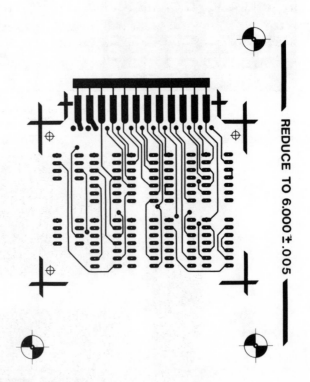

FIGURE 8-16B
Non-component
side traces
& pad master.

REDUCE TO 6.000 ±.005

Markings such as nomenclature, part number, and component symbols are frequently desired on multilayer P.C. boards. Such markings can be made either by etching simultaneously with the conductor pattern or by printing in ink.

Etched markings are usually lower in cost, provided the marks are large enough. Care must be taken when locating these marks so as to prevent any possible shorts between traces or pads on the final board.

Markings printed by using epoxy-based ink may be smaller than etched markings. Small printed markings will be easier to read, and the ink is as good an insulator as the base laminate. Also, ink markings remain readable after flow soldering, since solder will not adhere to the ink.

Whatever method of identification is used, the marks should be placed on outer layers of the final product.

HOLE SIZES

A plated-through hole must be sufficiently large enough to allow the required deposit of metal on the wall. (See standard hole formula, page 59.) As a rule of thumb, minimum hole diameters should be 1/2 of the board thickness after plating. This will allow the deposition of approximately .0005-in. to .0002-in. of metal on the wall of each hole. The metal deposited upon the surface will be up to two times as much as deposited in the hole. Very small holes can be plated-through. However, the ratio of metal deposited on the surface to that deposited within the holes increases.

Round holes should be used in multilayer boards unless a round hole is completely unacceptable because of the shape of some components to be mounted on the P.C. board. Slots, square holes, etc., add significantly to the cost of any P.C. board because of the complexity of the operation. Therefore, try to avoid slots or square holes. Figure 8-17A&B illustrates the cross section of a plated-through hole of a multilayer P.C. board.

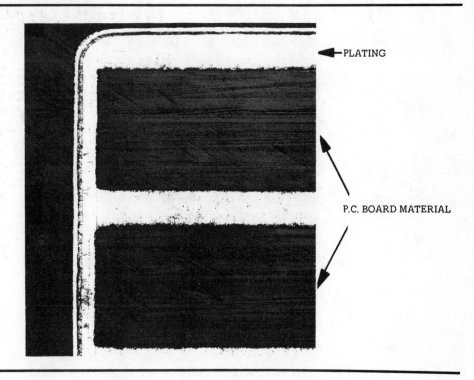

PLATING

P.C. BOARD MATERIAL

FIGURE 8-17A
Cross section of a
plated-through hole
of a multilayer P.C.
board (micro-
section).

3/4 of board thickness
This is a good distance between
the two edges of the hole.

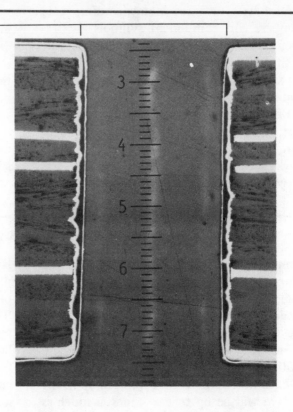

FIGURE 8-17B
Cross section of a
plated-through hole
of a multilayer
board.

HOLE
LOCATION

There are three methods of locating holes in multilayer boards:

1. Hole Location by Pad Center on the Reduced Artmaster (or Film): Using this method, all holes are located with reference to pad centers. This method permits the use of optical drilling with holes located directly from the artmaster (or film). Figure 8-18.

2. Hole Location by Dimensions from the Tooling Holes: This method of locating hole centers is by dimension from Datum lines or X and Y coordinates. When this method is used, the artwork must be carefully laid out to match dimensions which identify hole locations. Figure 8-19.

   Groups of holes may be required to match a specific component pattern such as an IC, TO-5 case or transistor pattern (Figure 8-20). These holes may be located dimensionally. In locating this type of group of hole one hole within the group should b designated as a reference hole and l cated dimensionally to the toolin hole. The remainder of the holes witl in the group should be located dimer sionally to the reference hole.

3. Hole Location by Grid Pattern: Mo designers locate hole centers on a pr determined grid pattern. Grid pattern containing increments of .100-in., an .050-in. are commonly used. Liste below are some of the reasons fo using a grid.

   a. Provides a standard base for spacir and locating component mountir holes, resulting in a simplificatio of component lead bending whe machine bending and trimming used.

   b. Using a grid pattern on the taping helpful in precisely locating pads o a "pattern to hole" and layer to layer dimensional system (see Figure 8-21).

c. A grid system is often used only for design standardization. Unless a good reason for locating holes to a grid exists, all holes should be located to pad centers. Multilayer boards or any type of P.C. board with holes located to pad center will have a better registration and cost less to produce, as one potential set of tolerance build-up is eliminated.

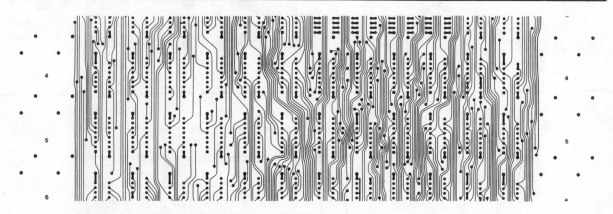

FIGURE 8-18
ocating from pad
ters for drilling.

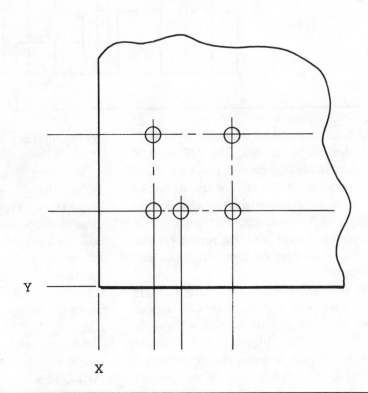

FIGURE 8-19

FIGURE 8-20

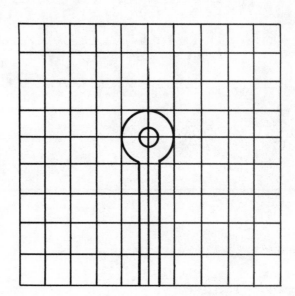

FIGURE 8-21

MATERIALS

The base material of the multilayer P.C. board should be specified as GE or GF grade glass-laminated epoxy, polyester film, polyimide film, or teflon. The use of epoxy glass as a base material will assure high stability and minimum warpage. This information should also be noted on the fabrication drawing or company standard specification.

The individual inner and outer layers should have one (1) or two (2) ounces copper clad or a minimum thickness of .001-.0025 inches (Figure 8-22). Outside layers should be one ounce copper since the plating adds to the external copper and makes etching more difficult. Photo # shows individual inner and outer layers.

The etched layer will be laminated together using prepreg or "B" stage epoxy glass cloth. Prepreg or "B" stage epoxy glass is a single layer of glass cloth which has been saturated with a specific amount of epoxy resin. The resin is uncured or partially cured, and will become plastic and flow under the application of heat and pressure. Photo #2 shows "B" stage material.

The prepreg or "B" stage material will range in thickness from .0025-.006 of an inch before lamination. The thickness used

MULTILAYER P.C. BOARDS

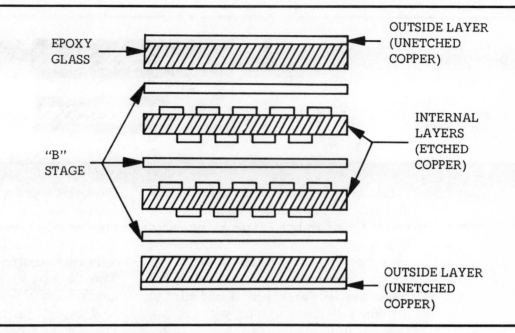

EPOXY GLASS

OUTSIDE LAYER (UNETCHED COPPER)

"B" STAGE

INTERNAL LAYERS (ETCHED COPPER)

OUTSIDE LAYER (UNETCHED COPPER)

FIGURE 8-22

Photo # 1
Inner and outer layers

Photo # 2
"B" Stage material

will depend upon the thickness of copper on each layer and the number of layers to be laminated together. Therefore, it is recommended that the final cross sectional thickness of the multilayer board be specified on the fabrication drawing (Figure 8-23) and the manufacturer be permitted the latitude of selecting laminating materials which will produce the desired structure.

THICKNESS
±0.007

LAYER 1
LAYER 2
LAYER 3
LAYER X

FIGURE 8-23

LAMINATION

Laminating is one of the most critical processes in producing a multilayer P.C. board. A multilayer P.C. board is produced from a combination of etched internal layers and prepreg or "B" stage material. Figure 8-24 shows a typical multilayer board lamination lay-up process. Lamination is performed by using fixtures with registration pins to maintain etched layer relationships to one another. Once the layers are registered, the multilayer assembly is then placed under high pressure and heat, usually 300 PSI and 350°F for about 55 minutes. The application of heat and pressure to the laminate fixture package converts "B" stage semi-cured material between etched layers to fully cured "C" stage epoxy. During the process, the "B" stage resin becomes a liquid adhesive which fills all etched circuit pattern voids and bonds the layers together. The structure must be totally free from all voids and this can be done only by a complete adherence of the epoxy resin throughout the conductor patterns and glass fibers on the inner layers. Two (2) ounce copper, for example, requires a larger amount of resin to fill the voids around the conductor than does one (1) ounce copper. Similarly, three (3) ounces copper require more.

The bonding materials, which are obtained from different suppliers, differ considerably from one another in thickness, having an effect on the end product thickness. Therefore it is recommended that the manufacturer of the multilayer P.C. board be given the freedom to select the base and prepreg or "B" stage materials that will meet the required finished thickness specified on the fabrication drawing.

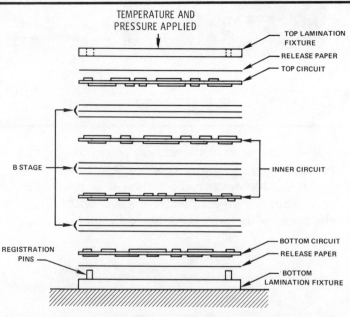

FIGURE 8-24
Typical multilayer
board lamination
lay-up process.

TEMPERATURE AND
PRESSURE APPLIED

TOP LAMINATION FIXTURE
RELEASE PAPER
TOP CIRCUIT

B STAGE

INNER CIRCUIT

REGISTRATION PINS

BOTTOM CIRCUIT
RELEASE PAPER
BOTTOM LAMINATION FIXTURE

The following photos show the board lamination lay-up process:

Step #1: Shows bottom lamination fixture with release paper.

Step #2: Shows registration pins, bottom circuit, and "B" stage material.

FIGURE 8-25

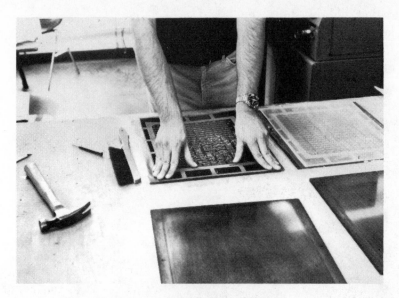

Step #3: Shows top circuit and top lamination fixture.

Step #4: Shows hot press for the application of heat and pressure to laminate the board.

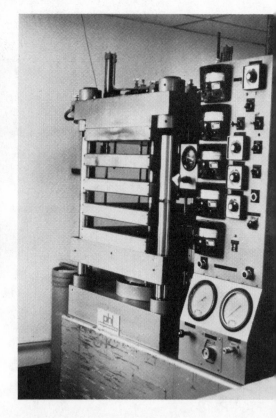

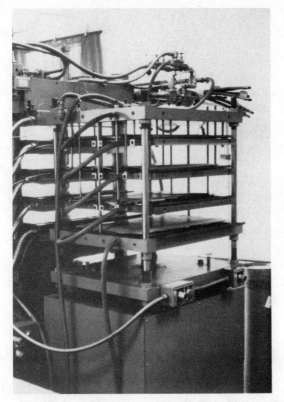

Step #5: Shows cold press. After the board is unloaded from the hot press, it may be desirable to quickly transfer the board to another press for cooling under pressure. Cooling under pressure is a good practice to minimize warping of the finished laminate.

After final unloading of the cooled board from the laminating plates, some advocate the use of a post-cure or bake cycle. Such a cycle normally consists of placing the board between metal plates to maintain flatness and loading the board into an oven for approximately one hour.

Step #6: Most multilayer board manufacturers check the completed board for material movement, shrinkage, growth, misregistration, voids, delamination, blistering, and a host of other flaws.

STEP 7A: Multilayer board manufacturers use a lab to check the completed board for material movement, shrinkage, growth, misregistration, voids, delamination, blistering, and other potential flaws. In the quality-control room, random samples are taken. The P.C. board is cut in half through the middle of a hole. It is then placed vertically in a molding cup and resin poured in, level with the cut hole. After the resin hardens, the top of the cup is lapped to a very smooth surface. The cut hole is then inspected microscopically for defects.

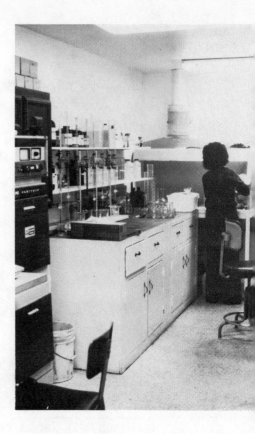

PLATING OF HOLES AND CONDUCTORS

Copper is the first metal deposited on the walls of a plated-through hole in a multilayer P.C. board. The plating process used for the multilayer P.C. board is the same as for a double-sided P.C. board.

Although some P.C. boards are produced with only copper-plated circuit patterns, the majority of P.C. boards have conductors plated with a second metal or alloy. In many instances, these conductor patterns are plated with tin-lead for easy assembly soldering. Gold is plated on some conductor patterns, but is most commonly used for connector fingers or conductor terminations. Tin-nickel and nickel-rhodium alloys are used where excessive wear may occur on the P.C. board.

The minimum recommended plating thickness is from 80 millionths of an inch for gold to 300 millionths of an inch for tin-lead, tin-nickel, and others.

CUTTING TO FINAL SIZE AND SHAPE

To facilitate production and handling, one sheet of cooper may actually contain many final etched conductor patterns. Figure 8-26 shows a typical multi-etched sheet.

Various methods are used to cut multilayer P.C. boards to their final size and shape. For small to medium quantities, routing to size and shape is the most common, and parts thus produced will be within ±.010 inch of the specified dimension. If large quantities are required (for example, a P.C. board used in the common calculator or digital watches) die and special tooling are used to produce the final shape. The technique is more expensive, although the final dimension is very accurate and consistent. Cost is then justified by quantity.

FIGURE 8-26

P.C. BOARD THICKNESS

On multilayer boards designed with fingers or edges intended for insertion into connectors, the board thickness is important. To meet the requirements of most commercial connectors, the P.C. bo[ard] thickness typically is .060 and ±.007-in. the nominal thickness specified. This standard tolerancing. See Figure 8-23.

FABRICATION DRAWING

Possibly the most important document required for an accurate and economical production of a multilayer P.C. board is the initial preparation of a correct and complete fabrication drawing.

The following list of items should be required on any fabrication:

1. An outline drawing of the finished multilayer P.C. board that shows all the necessary dimensions together with their tolerance.
2. A cross-sectional view showing the desired end product, indicating all layers (by number), and critical dimensions (Figure 8-23).
3. Conductor pattern images for non-component side identified by layer number (this is optional but very desirable). Also, each conductor pattern layer should be identified numerically.
4. A description of the base material (G10 or G10FR) or other selected materials.
5. A table showing hole diameters, tolerances, and number of holes, and which holes are clearance or plated-through holes. Also, a coding syst[em] that shows which holes in the ta[ble] can be identified on the multila[yer] circuit board. See Figure 8-27 fo[r] typical hole table.
6. A system to locate holes either [on a] standard grid pattern, or by X-Y [co]ordinate dimensions. Use only [one] corner when specifying coordin[ate] dimensions.
7. Show any markings such as refere[nce] designations, nomenclature, part nu[m]ber of the fabricated P.C. board, p[art] number of the board assembly, and [so] forth. These markings can either [be] printed on the board with special i[nk] or etched simultaneously with [the] conductor pattern. In the latter ca[se] these markings will appear on the a[rt] work.
8. Name and number of any other no[tes] and specifications that may apply [to] the end product.

Hole	Quantity	Diameter	Tolerance
A	2	.375	± 005
B	13	.093	± 002
C	7	.125	± 003
D	All other	.040	± 003

FIGURE 8-27

P.C. BOARD CHECKLIST

One of the last things done before a multilayer P.C. board is released for fabrication is the checking of all required design and artwork parameters. The following is a typical list that could be used for this important final step in the multilayer design:

1. Part numbers and logo.
2. Required reference designation. All IC's, transistors, pots, conn. numbers, TP, E numbers.
3. Rev. letter.
4. Title Block—complete.
5. Reduction targets.
6. All pin 1 on IC's are marked, positive end of capacitors marked, cathode end of diodes marked.
7. Board thickness.
8. Environmental and performance requirements.
9. Made in USA label.
10. Size of board and its relationship with mating structures. .125 min. clearance on two sides from all components.
11. Connector location on the P.C. board and its relationship with mating connector.
12. Operating temperature.
13. Conductor width: The standard width at 1:1 shall be .025 for signal lines and .050 for voltage, (or company standards).
14. Conductor spacing at 1:1 shall be .025 standard, .020 min., or company standards.
15. Component lead spacing.
16. All pads must be on grid intersections except feed-through pads.
17. Rough assembly drawing.
18. Parts list.
19. Ground stakes should be placed on the P.C. board approx. every 16 sq. inches.

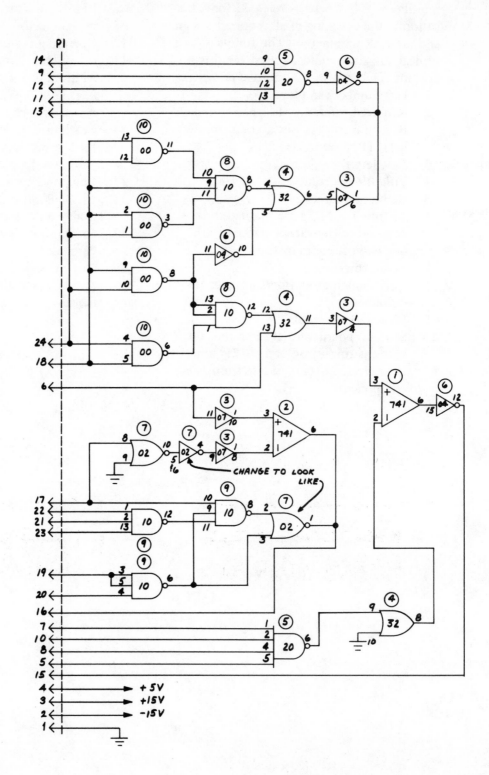

Schematic shown is
for tape-up shown
in Figure 8-16A & B

Chapter

9

BOARD OUTLINE

Before a printed circuit board is started (locating the components and traces), several items must be defined. One of the first is the working board area; as shown in Figure 9-1.

The following items are some of the considerations that should be involved prior to starting the actual circuit packaging:

1. What type of housing or enclosure will the completed board assembly be mounted in?

2. What method of fastening or mounting the board is best for this application (card guides, screws, standoffs, etc.)?

3. What type of electrical interface is available (connector, cable, wires, etc.)?

4. Will the board require card extractors or special extractor tools to aid in removal of board from enclosure?

5. If board is large or irregular in shape, are stiffeners or special braces required?

These questions and others that are typical to any number of design requirements are covered in the following outline.

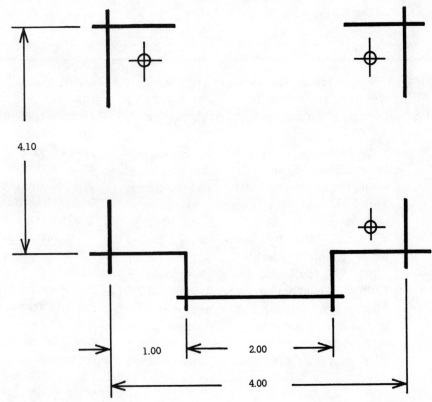

FIGURE 9-1
Board Outline   NOTE: All dimensions ±.02-in.

BOARD OUTLINE Depending on how the P.C. card mounts or slides into the next assembly, the configuration of the board must be such that it clears any obstacles or interferences. Therefore, a typical board may have cutouts, radiuses or other irregular shapes. It is important to remember that every cut, hole or routing operation that is used increases the cost of the board. To minimize cost, keep the configuration as close to a basic square or rectangle as possible.

METHOD OF MOUNTING If the type of electrical interface requires the P.C. card to insert into a mating connector, then card guides offer the most convenient solution. They not only provide a quick "connect/disconnect" capability, but also provide a means of testing the board out of the unit by means of extender cards. One of the most economical ways to provide this type of interface is by the use of an edge connector. This method not only eliminates the cost of buying an additional connector, but saves manufacturing time by not having to assemble that second connector.

If the electrical interface is by means other than a connector or a card guide is not practical, then mounting holes must be included on the board so that it can be installed with screws, standoffs or other mechanical fasteners.

Whatever means is decided upon, sufficient clearance should be provided so that components or conductors will not interfere or short out to the mounting hardware.

WORKING BOARD AREA Regardless of the board configuration, a few basic rules must be considered prior to beginning the component packaging phase.

1. *Tolerance* - All board outlines must be expressed to include some tolerance. Most common are ±.010- and ±.020-in. This is important because the actual working board area must be the smallest dimension that the board would be at worst tolerance condition. In the case of Figure 9-2, the board dimensions would be 8.08 x 4.03 inches or 32.56 square inches. That .02-in. tolerance may not seem like much; but if designing a very dense board where a trace runs within .040-in. of board edge, that tolerance could be of significant value.

2. *Component Interference* - If the P.C. board mounts into an enclosure that has an interference or obstacle near the board area, careful planning must insure that the P.C. board assembly (including components and hardware) will not have mechanical or electrical interference. In the case of some critical R.F. circuits, the electrical engineer must approve the relationship of the components on the P.C. board to the environment that they will be packaged at the final assembly. Closeness to the sheet metal chassis or other circuits could alter the function of the circuit being packaged.

Component interference is also a major consideration within the P.C. board. All components, heat sinks, fasteners, etc. must be considered on the layout at their *maximum size* or greatest area of interference. This is especially critical in considering such components as special designed transformers, chokes, etc., because the prototype device that may be provided as a sample sometimes grows in size between

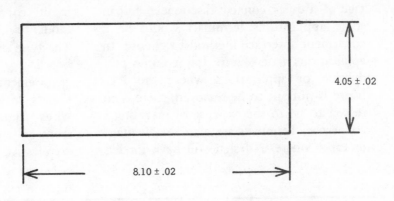

4.05 ± .02

8.10 ± .02

FIGURE 9-2
Working Board
Area

the prototype stage and the final production of that particular part. If there is no data sheet or reliable specification on any given part or component, leave adequate space to include maximum interference area of that part.

3. *Method of Mounting Board at Next Assembly* - There are various methods available to secure the board assembly to the next assembly. The application best suited for the particular board must be decided prior to starting board layout. This is important because if standoffs are used to mount the board, make sure that components or traces will not be in the position where a hole is drilled or a fastener is installed. In the case of card guides, sufficient area would have to be left clear to allow room for the card guide along the edge of the board.

Other items associated with the board mounting are items such as: Card ejectors,

mounting holes, stiffeners, and metallic or non-metallic card guides (If card guides are non-metallic, traces may be routed right out to the edge of the board).

One basic consideration on *all* printed circuit packaging is the leads that protrude through the non-component side of the printed circuit board. Adequate clearance must be provided for all leads, wires, hardware, etc. to ensure they will not short out to an adjacent board, sheet metal or object mounted underneath the board. The allowance for these leads is generally not to exceed .060- to .100-in.; therefore, a minimum clearance of an additional .060- to .100-in. should be designed into the mechanical package. If the board is an exceptionally large board or made of thinner material, it might require even additional clearance to ensure that flexing of the board under vibration or stress would not allow the leads to short out.

INTERFACE CONNEC-TIONS

One of the basic requirements of any P.C. board is to provide means of electrical interface between the board and its associated equipment. The most common method of accomplishing this is by the use of connectors, terminals and/or cables.

1. *Connectors* - The type of connector or interface that will be used for any particular board is generally decided upon during the mechanical design of any given piece of

equipment. This is necessary because not only does the P.C. board have to consider the type of connector or connection, but the associated mechanical package must provide space for the connector or connection that will mate with the printed circuit board. The choice of connectors that are available are too numerous to cover, but there are models and configurations to fit almost any requirement.

2. *Terminals* - Connectors have the advantage of a quick connect-disconnect, but in some applications terminals would be a convenient interface method to choose. In applications that require low volume production or applications where the P.C. board is not apt to be removed, there is no reason to go to the expense of installing a connector. Terminals are also an advantage in cases where a design will have models that are the same basic circuit but require input connectors to be different from model to model.

The type of cable best suited for any given board is greatly dependent on the electrical requirements and/or contract mandate. Cables, in general, range from a few single wires (bundled together and connected from unit to unit) to cable assemblies that are massive in diameter.

PRINTED CIRCUIT BOARD MATERIALS

Materials for P.C. boards generally fall within a couple of standard laminate grades with "G-10" and "FR-4" being the most common due to the availability of the material. Table 9-1 lists these laminates.

Although the laminates range in a variety of material, the finished P.C. board material (laminate with copper applied) is generally treated the same as far as etching the final printed circuit board.

Grade	Composition	Remarks
XXXPC	paper / phenolic	High moisture resistance
FR-2	paper / phenolic	Similar to XXXPC but flame retardant
XXXP	paper / phenolic	Best mechanical characteristics of paper / phenolic grades
FR-3	paper / epoxy	High mechanical and electrical characteristics, flame retardant
FR-4	glass / epoxy	Flame retardant, chemical resistant, low water absorption
G-3	glass / phenolic	High flexural strength and dimensional stability
G-5	glass / melamine	High resistance, high impact strength
G-9	glass / melamine	Same as G-5 but better electrical characteristics
G-10	glass / epoxy	Same as FR-4 but not flame retardant
G-11	glass / epoxy	Same as G-10 but higher flexural strength under heat
G-30	glass / polymide	High demensional stability under heat, flame retardant
FR-5	glass / epoxy	Same as G-11 but flame retardant
GPO-1	glass / polyester	General purpose mechanical and electrical grade
GPO-2	glass / polyester	Similar to GPO-1 but lower flammability

TABLE 9-1 Major Laminate Grades Used by P.C. Suppliers

TOOLING HOLES

Tooling holes are used as P.C. Board reference points upon which other dimensions are based. These are noted in Figure 9-3.

During the manufacturing process, the manufacturer will "tool-up" or design his fixtures for drilling the P.C. board using these tooling holes as a reference point. It is, therefore, necessary that the dimensions in reference to the tooling holes be located accurately per the dimensions given on the fabrication drawing.

In cases of very dense boards, there may not be sufficient room on the board to locate the pads for the tooling holes; therefore, it will be necessary to locate them outside the perimeter of the board outline. They will appear on the final positive and negative film which the manufacturer will use to produce the board. The same accurate dimensioning would still apply, but after the board is completed, they would be trimmed away with the other excess material.

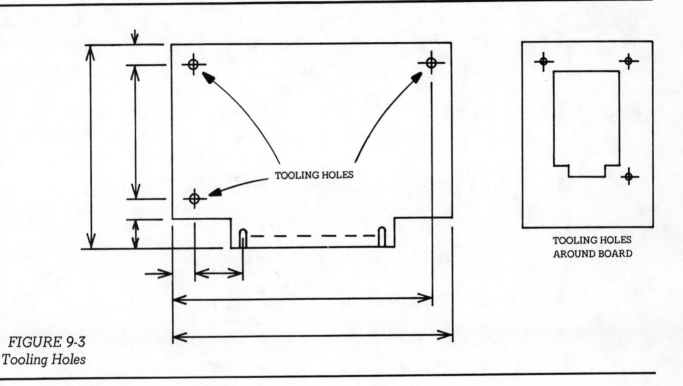

TOOLING HOLES

TOOLING HOLES
AROUND BOARD

FIGURE 9-3
Tooling Holes

Chapter

10

PRODUCTION CONSIDERATIONS

MECHANICAL INSERTION

This chapter introduces robot manufacturing and the pc board design considerations to accommodate it.

PC boards are often manufactured with the help of automatic insertion machines—industrial robots which are able to load a pc board with electronic components many times faster than is humanly possible. It is crucial for the pc board designer to understand how these machines work, and how an apparently arbitrary component layout can spell the difference between profit and loss during factory operations.

Robots, also called "board-stuffing machines" accept all the individual components of a pc board along with the empty board itself. The empty board has previously been processed through the conductive-cladding and hole-drilling stages, and is ready to be "stuffed" with electronic components.

Discrete components are supplied to the robot on reels; integrated circuits are supplied in tubes. For discrete parts such as transistors, diodes and resistors, the robot is programmed to cut one component from the supply reel, trim and shape its leads, orient the pc board with respect to the component, insert the component's leads through the proper holes, then cut and crimp the leads on the bottom side of the board . . . and then repeat the entire sequence for the next component.

Components are sequenced on the supply reel according to the robot's programmed sequences. It is the responsibility of the pc board designer to assure minimum board movement between each component insertion, since each millisecond needed for the robot to re-position the board subtracts directly from operating profit.

To that end, this chapter describes the design tradeoffs to be made, and offers quantitative evaluations for the most popular component configurations. This chapter also discusses machine travel-time, insertion-head changes and the limits of design acceptability.

BEST COMPONENT ARRANGEMENT

The most economical component arrangement for mechanized assembly is one in which all components are identical, all lead spans are the same, and all component bodies are oriented in the same direction. This arrangement is the lowest cost method because components can be fed directly from a supplier's tape reel to a standard insertion head and directly into the P.C. board which has only to be put into the machine and taken out once.

However, the component arrangement that is best for automatic assembly is rarely the best from a functional design standpoint, so most P.C. boards are a compromise of these two needs.

One of the best component arrangements is shown in Figure 10-1, where all lead spans are the same and all the components cases (or bodies) are oriented in the same direction. This arrangement will give an assembly cost factor of 1.00, based on insertion costs with a machine having a manually-variable center-distance insertion head. Sequencing costs are included because few P.C. boards have components that are all electrically alike.

Figure 10-2 is a compromise arrangement. The different components still are oriented in the same direction, but the sequenced parts have two or more different lead spans. This difference increases the machine assembly cost factor to 1.05. This is still an acceptable cost compromise.

Another compromise arrangement is Figure 10-3 which has two or more different lead spans and six of the components are oriented 90 degrees to the others. Having two orientations increases the cost factor to 1.25. This is also an acceptable compromise.

In the compromise of Figure 10-4, seven of the components are oriented 90 degrees to the others, and the components in each group are scattered around the P.C. board. Also, each group has three different lead spans which adds to the cost. The cost factor here is 1.35; this may be very close to an undesirable compromise.

This could be undesirable because of the excessive machine traverse time required to

insert similarly oriented components th are scattered around the P.C. board. T most economical use of the automat machine dictates that all components wi the same orientation be inserted before t board is re-positioned for another orient tion. In this case, the machine has to stopped six times to manually change t insertion head for each of the six differe lead spans. However, even with the sca tered components, this layout would acceptable with a cost factor of 1.35.

The arrangement of Figure 10-5 is t tally impractical; there are too many diffe ent component orientations which result excessive idle machine time while the boa is being realigned to the insertion head f each of the orientations. The requireme to hold the board at a number of differe angles to the insertion head increases t tooling cost. Cost factors of the example Figures 10-1 through 10-5 have assume insertion with a manually variable hea Actually, many insertion machines are use today, not only fixed heads, but hea that can be switched back and forth b tween two lead spans. Effective use these machines requires even closer evalu tion of component arrangement.

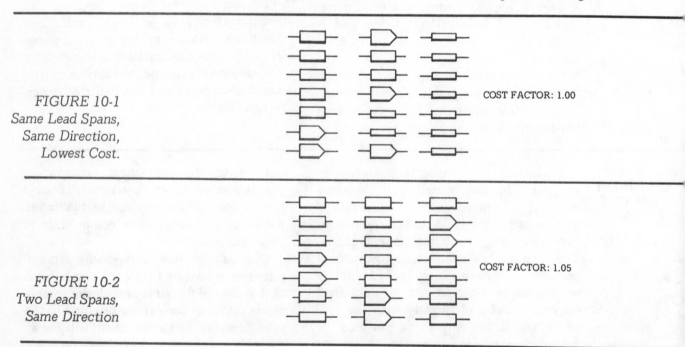

FIGURE 10-1
Same Lead Spans,
Same Direction,
Lowest Cost.

COST FACTOR: 1.00

FIGURE 10-2
Two Lead Spans,
Same Direction

COST FACTOR: 1.05

PRODUCTION CONSIDERATIONS

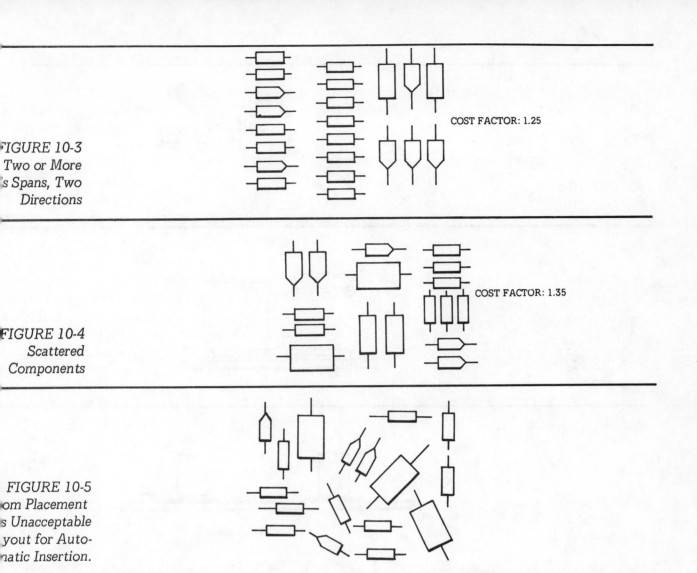

FIGURE 10-3
Two or More
Spans, Two
Directions

COST FACTOR: 1.25

FIGURE 10-4
Scattered
Components

COST FACTOR: 1.35

FIGURE 10-5
Random Placement
Is Unacceptable
Layout for Auto-
matic Insertion.

AXIAL COMPONENT ARRANGE-MENTS

One approach to automatically insert transistors with radial leads is the axial method shown in Figure 10-6. Transistors inserted in this fashion are sequenced on tape and inserted with a machine having a modified axial-lead-component head.

The best arrangement for axially inserted transistors in a specific sequence is uniform lead span and orientation, such as Figure 10-7.

By pre-forming leads as shown in Figure 10-7, components with radial leads can be processed like those with axial leads.

Once the transistor is on the tape, the insertion problem again becomes one of compatiable wire diameter, since it is handled only by its leads, and a body clearance to a maximum size. The results, nearly all vendors' components can be used by a single insertion head without re-tooling. See Figure 10-8. Spans from .300 to .400 are possible.

RADIAL COMPONENT ARRANGE-MENTS

For radial lead transistors, the device is held by its body while its radial leads are inserted straight into the board after the leads have been trimmed to a suitable length. After insertion, the leads are crimped, two one way and the third the other. Figures 10-10 through 10-13 show radial lead transistor arrangements. An

PRODUCTION CONSIDERATIONS 163

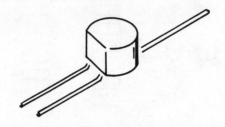

FIGURE 10-6

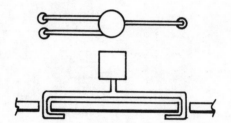

FIGURE 10-7

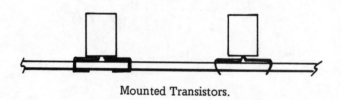

Mounted Transistors.

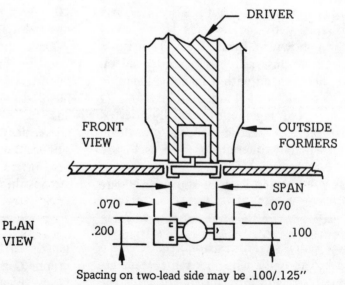

FIGURE 10-8

Spacing on two-lead side may be .100/.125″

Transistor Assembly Tools.

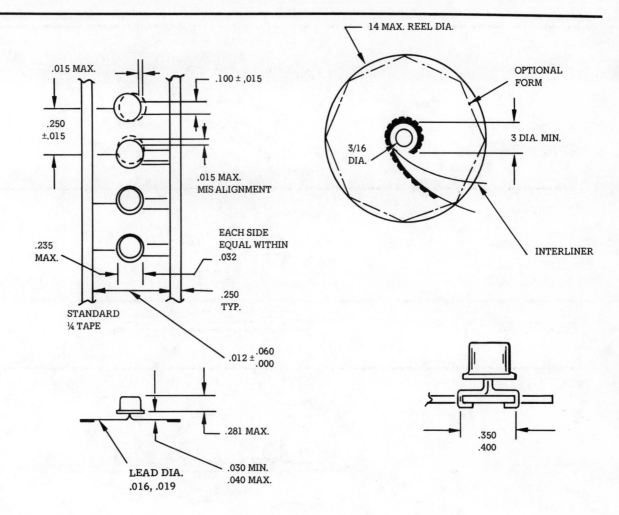

FIGURE 10-9

OR MACHINE INSERTION OF TO-18, TO-92, TO-98 AND
THER SIMILAR TRANSISTORS.

angle not divisible by 90 is unacceptable in radial insertion, and in axial lead assembly.

A little forethought in design and manufacturing of printed circuit boards makes a big difference in the cost of the assembly. Designers usually have no direct control over the assembly method or insertion machine used in production, but by choosing the simplest and most feasible component arrangement, this frees the manufac-

turing engineer to select the most economical method.

Designers need to be aware of the relationship of the board size, hole sizes, insertable area and component spacing to the requirements of the automatic insertion machine. If possible, the designer should try to limit the printed circuit board design on component orientation. Also, try to keep components with the same orientation grouped together.

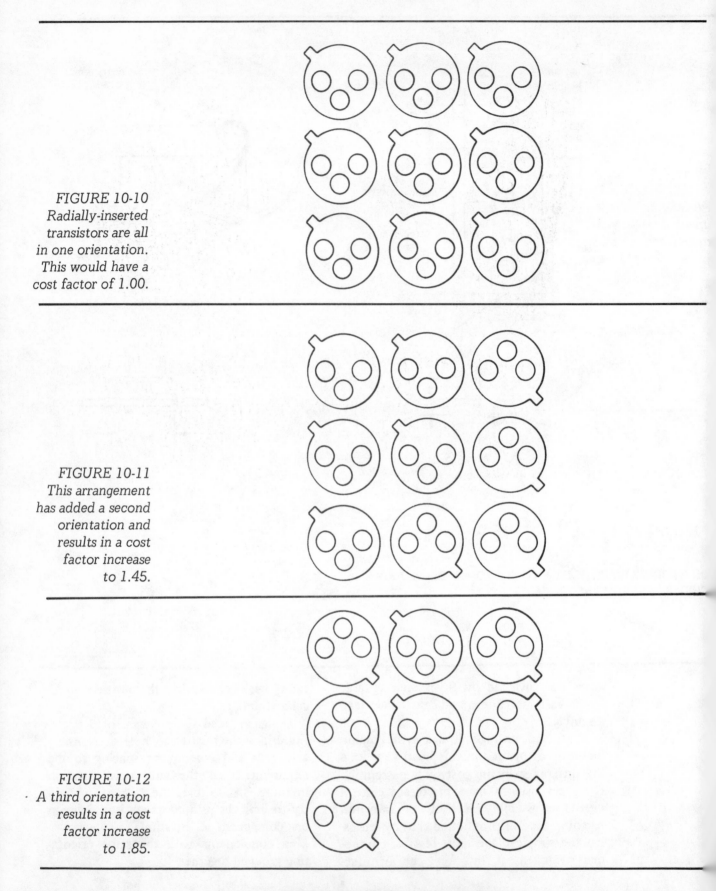

FIGURE 10-10
Radially-inserted
transistors are all
in one orientation.
This would have a
cost factor of 1.00.

FIGURE 10-11
This arrangement
has added a second
orientation and
results in a cost
factor increase
to 1.45.

FIGURE 10-12
· *A third orientation*
results in a cost
factor increase
to 1.85.

FIGURE 10-13A
A random arrange-
ment. This arrange-
ent has more than
doubled the cost
factor to 2.25.

TOOLING HOLES To obtain the maximum accuracy for P.C. boards with respect to where holes are located, tooling holes should be used, with the longest possible span between them on the same edge if possible. These tooling holes will establish the vertical and horizontal axes for fabrication. See Figure 10-14.

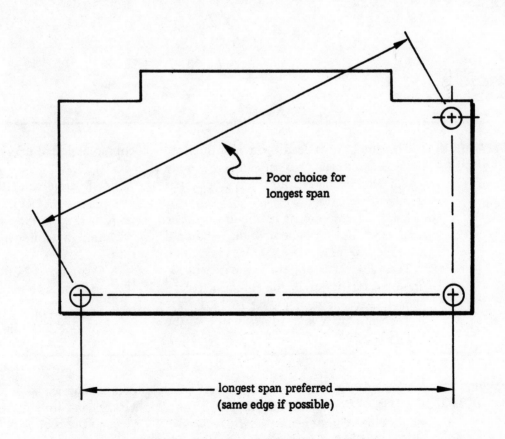

Poor choice for longest span

longest span preferred
(same edge if possible)

FIGURE 10-14

TOOLING HOLE TOLERANCE & SIZE

Hole size
Recommend hole sizes =.100,125,150
Most Common=.125

Tolerance—The diameter of the tooling hole should be held to a very close toler-ance. Any shifting of the board due to a oversize tooling would directly affect th hole tolerance and size of the componer hole. See Figure 10-15.

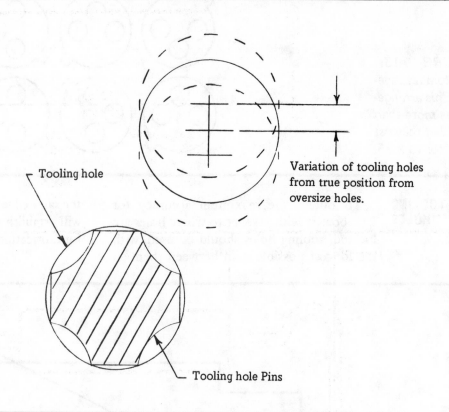

Variation of tooling holes from true position from oversize holes.

Tooling hole

Tooling hole Pins

FIGURE 10-15

SPAN CON-SIDERATIONS

The component lead span is influenced by several factors such as lead form, body length, lead diameter, and the type of insertion head being used.

Standard component configuration. The span used for component with a standard lead. Is .250 in. min. to 2.00 in. maximum.

The value of X for the standard configuration is a function of the component lead diameter. See example Figure 10-16.

The value of X & Y is a function of the component lead diameter. See example Figure 10-17.

For Example on How This Works
Let's use a RL07 resistor case whos body is .281 in long and a wire diameter c .027 in., the following min. spans are pos sible.

A. Standard= 2x.075 + .281 =.431=i span or .450 min.

B. Stress Loop= 2x.125 + .250 =.531 i span or .550 min.

COMPONENT HOLES

Finding the hole size necessary for automatic component insertion is a function of the component *"Lead Diameter,"* all "Machine Tolerances" for guiding the component leads, the table positioning ac curacy, and hole pattern accuracy with re spect to the tooling holes.

PRODUCTION CONSIDERATIONS

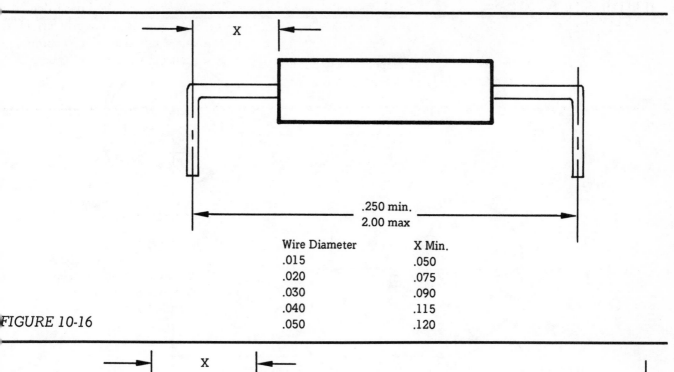

Wire Diameter	X Min.
.015	.050
.020	.075
.030	.090
.040	.115
.050	.120

FIGURE 10-16

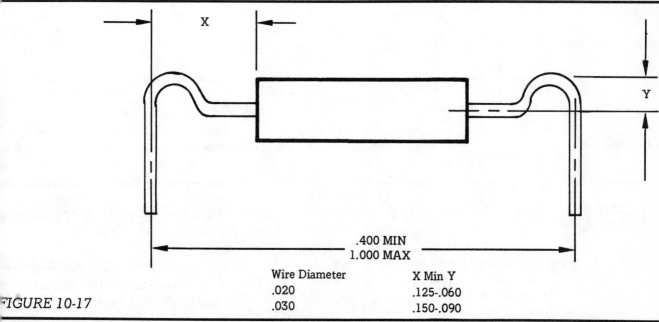

Wire Diameter	X Min Y
.020	.125-.060
.030	.150-.090

FIGURE 10-17

All of the mechanical tolerances allowed for P.C. board drilling may be expressed as the determination of the min. dia. of the component hole and min. tooling hole.

The *Minimum Component Hole Diameter* is determined from the following equation:

D1 = The maximum diameter of the board tooling holes minus the diameter of the tooling pin.

D2 = The total variation of the component hole position as permitted by the allowance tolerances of dimensions A and B of Figure 10-18.

LD = Component Maximum Lead Diameter

MA = The insertion machine accuracy—Total allowable variation.

MCHD = D1+D2+LD+Ma

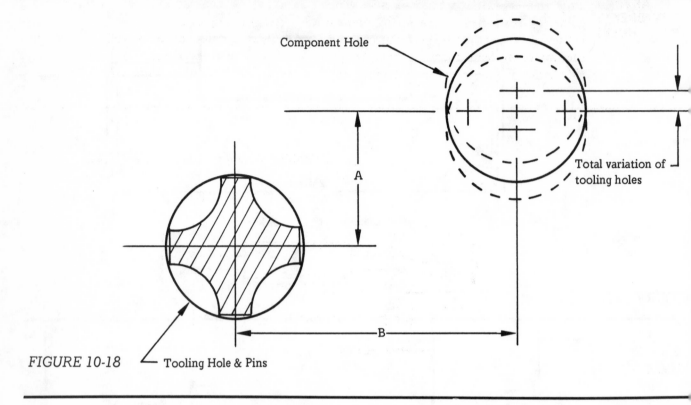

Component Hole

A

B

Total variation of tooling holes

FIGURE 10-18 — Tooling Hole & Pins

Example

Step A. Tooling hole diameter .125 in. + .002, - .000

B. Locating Pin diameter .125 in. + .000, - .001

C. Therefore from (A) and (B) above: D1= .003 Inch.

D. Hole position Tolerance: + or -.003 inch (TYPICAL)

E. Therefore from (D) above: D2 = .006 inch.

F. Component lead diameter: .025 inch + or - .001 inch

G. Therefore from F above: LD = .026 inch.

H. Machine accuracy: + or - .002 inch (TYPICAL)

I. Therefore from H above: MA = .004 inch.

J. Therefore from C, E, G, and I above: MCHD = D1 = D2 + LD + MA = .003 + .006 + .026 + .004 = .039 inch.

NOTES

1. For nearly all practical purposes the sum of all insertion machine factors including D1, and MA, can be considered to be .008 in. This reduces the equation to the total drilled hole position tolerance variation (D2) plus the maximum wire diameter (LD) plus .008 in.

2. In the event that the calculated minimum component hole diameter becomes too large with respect to the circuit requirement, other tolerances may be tightened to allow for the use of a smaller component hole.

PRODUCTION CONSIDERATIONS

CLEARANCE AROUND COMPONENTS HOLES

Axial lead component insertion heads are equipped with outside formers which guide the component leads to the point of insertion into the board; clearance around the insertion lead formers must be provided. See Figure 10-19 and Figure 10-20 for the clearances required between the lead being inserted and any adjacent component body or lead.

Typical value of X is .093 inch minus one-half of component lead diameter and a typical value of Y is .062 inch.

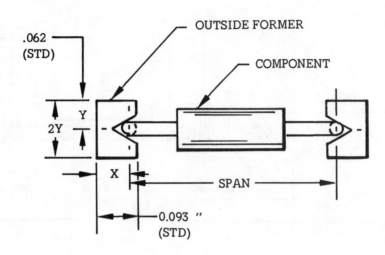

FIGURE 10-19

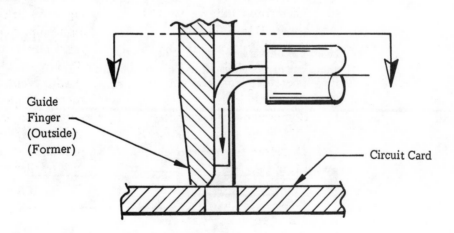

FIGURE 10-20

BODY CLEARANCE   The minimum clearance required around the body of the component will vary for different mounting configurations. The more common configurations are shown Figure 10-21, 22, 23, 24, 25, 26, a 10-27.

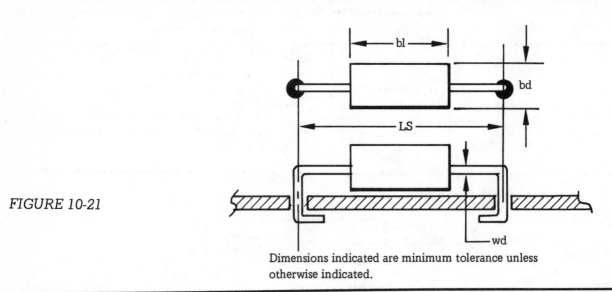

FIGURE 10-21

Dimensions indicated are minimum tolerance unless otherwise indicated.

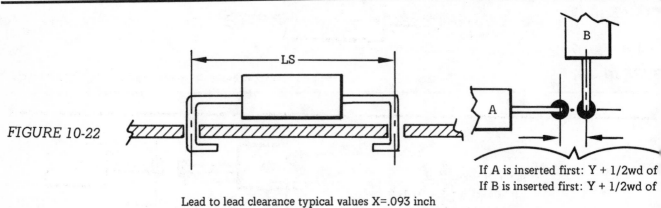

FIGURE 10-22

If A is inserted first: Y + 1/2wd of
If B is inserted first: Y + 1/2wd of

Lead to lead clearance typical values X=.093 inch minus one-half component lead dia. Y=.062 inch.

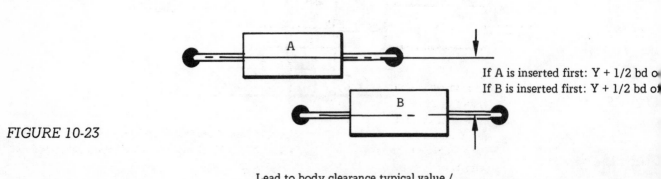

FIGURE 10-23

If A is inserted first: Y + 1/2 bd o
If B is inserted first: Y + 1/2 bd of

Lead to body clearance typical value /
Lead to body clearance typical value Y=.062 inch.

PRODUCTION CONSIDERATIONS

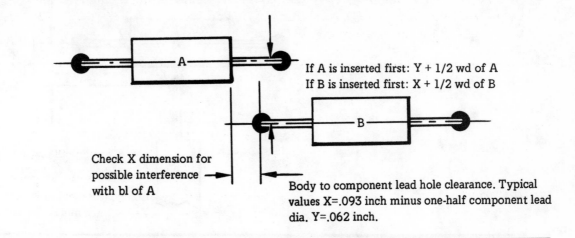

If A is inserted first: Y + 1/2 wd of A
If B is inserted first: X + 1/2 wd of B

Check X dimension for
possible interference →
with bl of A

Body to component lead hole clearance. Typical
values X=.093 inch minus one-half component lead
dia. Y=.062 inch.

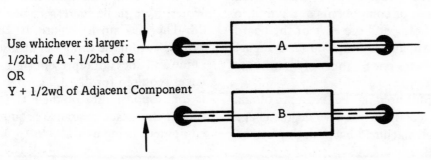

Use whichever is larger:
1/2bd of A + 1/2bd of B
OR
Y + 1/2wd of Adjacent Component

Body to Body Clearance Typical Value Y=.062 inch.

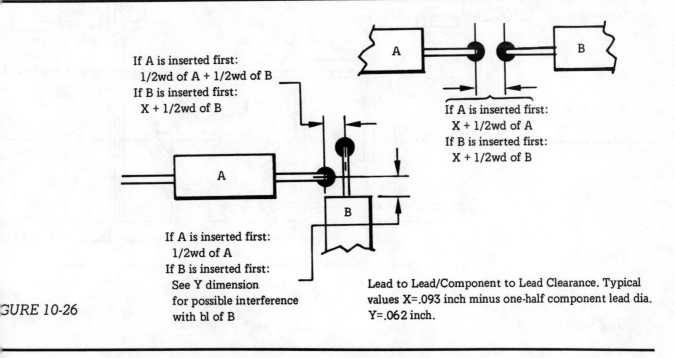

If A is inserted first:
    1/2wd of A + 1/2wd of B
If B is inserted first:
    X + 1/2wd of B

If A is inserted first:
    X + 1/2wd of A
If B is inserted first:
    X + 1/2wd of B

If A is inserted first:
    1/2wd of A
If B is inserted first:
    See Y dimension
    for possible interference
    with bl of B

Lead to Lead/Component to Lead Clearance. Typical
values X=.093 inch minus one-half component lead dia.
Y=.062 inch.

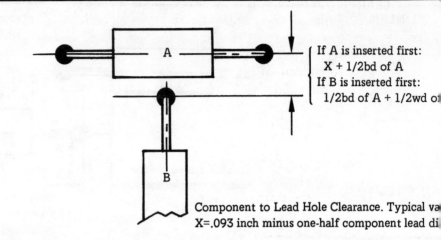

If A is inserted first:
X + 1/2bd of A
If B is inserted first:
1/2bd of A + 1/2wd o[

FIGURE 10-27

Component to Lead Hole Clearance. Typical va[
X=.093 inch minus one-half component lead di[

BELOW BOARD CLEARANCE

The board problems are most often the placement of components in relation to the tooling holes, and the edge of the board for areas which requires support. Figure 10-28 illustrates one method of modifying (by chamfering).

The board is holding the fixture to obtain minimum clearance. If the 45 degree chamfer were not machined back on the fixture,

the .175 inch and .375 inch dimensi[would have to be increased. See Figure 28. The diagram in Figure 10-28 show[typical automatic Component insertion r[chine and does not represent the exact Cle[ances required for "ALL" such machines[is the responsibility of the P.C. Designer[obtain the exact clearance required for [equipment being used.

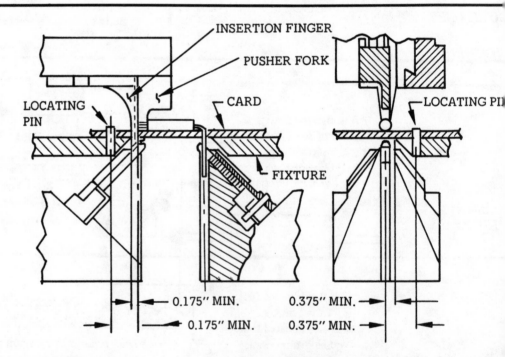

FIGURE 10-28

Board Holding Fixture (Typical Automatic Compon[
Insertion Machine)

PRODUCTION CONSIDERATIONS

LEAD CLINCHING

Lead forming is typically, along the centerline of the component and inward, back toward the body. The clinch is either 45 degree or 90 degree. See Figures 10-29, 10-30, and 10-31.

The length of the component lead (Clinch Length) is a function of the lead diameter, as shown in Figure 10-32.

Clinched Lead Length

* The clinch length is measured from the centerline of the component lead as it extends back.

* The minimum clinch length for small diameter leads are dependent on the hole size. To maintain a minimum clinch length, the hole size must not be more than .014 inch larger than the component lead diameter.

* 45 degree clinch length may be as small as .8 times the lead diameter This allows only .3 times the lead diameter extending over the terminal area.

Wire Dia.	Min. Clinch	Max. Clinch
.020	.030	.080
.030	.040	.080
.040	.050	.080
.050	.060	.080

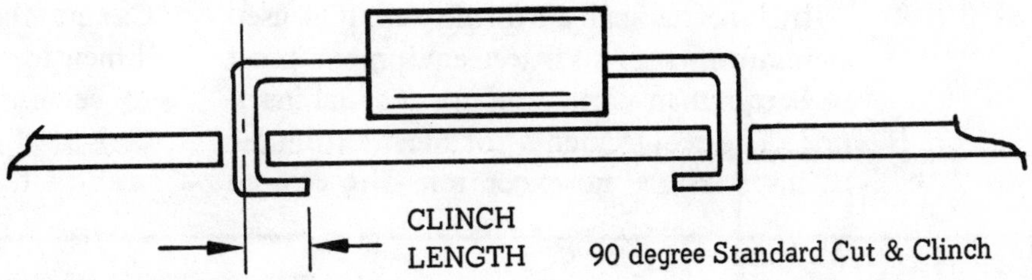

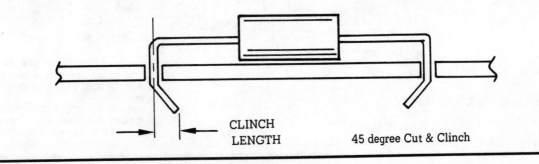

FIGURE 10-29   90 degree Standard Cut & Clinch

CLINCH LENGTH

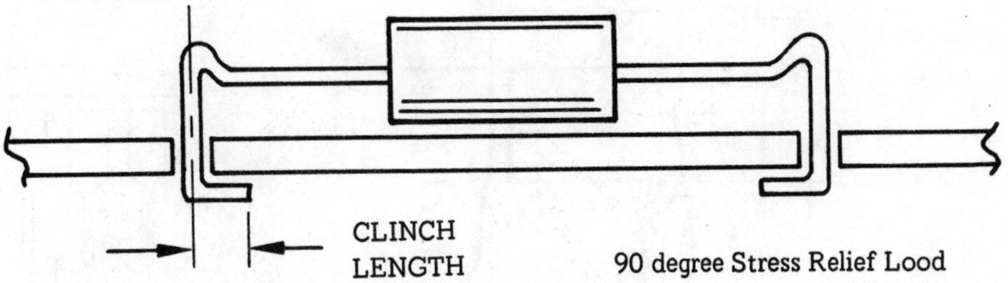

FIGURE 10-30   90 degree Stress Relief Lood

CLINCH LENGTH

FIGURE 10-31   45 degree Cut & Clinch

CLINCH LENGTH

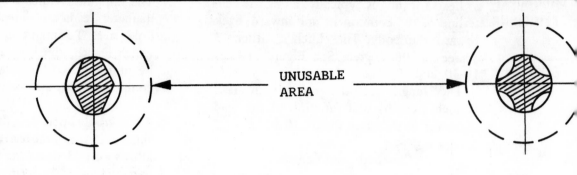

FIGURE 10-32

UNUSABLE AREA   As a general rule the minimum area around any tooling hole is approximately .375 inch radius, while edge locating and support methods usually require .250 along the guide edges.

IC CONFIGURA-TIONS   The mechanical uniformity of IC's used with automatic insertion equipment must be better than parts used for manual insertion. The requirements for most automatic IC inserters are no exception. The Printed Circuit Designer must obtain the deta dimensional requirements of the specifi to be inserted before proceeding with design. A commonly recommended cor uration for IC's is shown in Figure 10

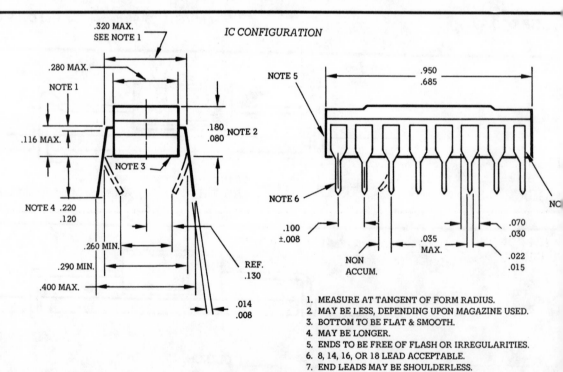

IC CONFIGURATION

1. MEASURE AT TANGENT OF FORM RADIUS.
2. MAY BE LESS, DEPENDING UPON MAGAZINE USED.
3. BOTTOM TO BE FLAT & SMOOTH.
4. MAY BE LONGER.
5. ENDS TO BE FREE OF FLASH OR IRREGULARITIES.
6. 8, 14, 16, OR 18 LEAD ACCEPTABLE.
7. END LEADS MAY BE SHOULDERLESS.

FIGURE 10-33

CARRIER-FED IC's Current methods of loading IC's into automatic insertion equipment is by means of plastic stick magazines. The tolerance on IC's required ofr this technique are shown in Figure 10-34.

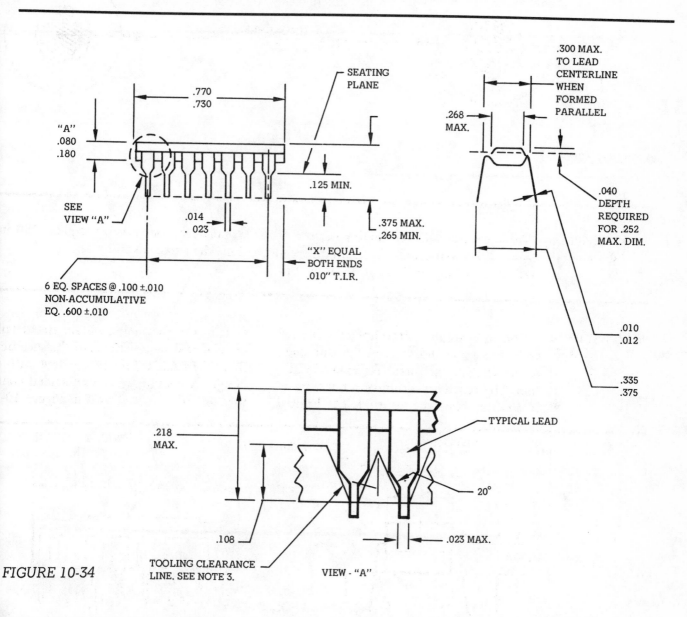

FIGURE 10-34

IC LOCATION CONSIDERATIONS It is recommended that all IC's are rotated in the same direction.

ARTMASTER TAPE-UP

IC SPACING   Figure 10-35 gives the end-to-end and side-by-side spacing requirements for most IC-insertion equipment. In general, if IC's and axial lead components are mixed on the same board, the same spacing ru[apply, providing the axial-lead compone[height is not greater than .250 in. at t[point of clearance.

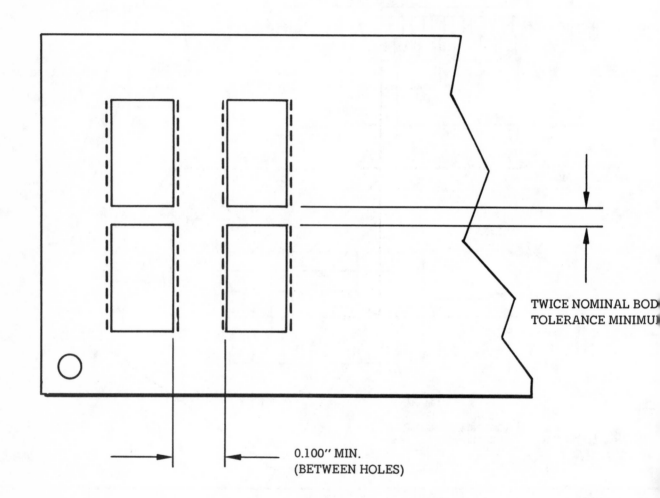

TWICE NOMINAL BOD[
TOLERANCE MINIMU[

0.100″ MIN.
(BETWEEN HOLES)

FIGURE 10-35
Recommended Min. IC Spacing

IC HOLE SIZE & TOLERANCE

In determining the minimum insertion hole diameter the lead must be considered as being the maximum effective diameter. The same formula for the hole size and tolerance applies as for round wires. Since there are fourteen to forty leads to be considered, the probability of worst case is increased.

For IC's the minimum insertion hole diameter must be calculated from the maximum diagonal dimension across the rectangular cross section of the leads. Each manufacturer could have slightly different dimensions and tolerances. See Figure 10-36.

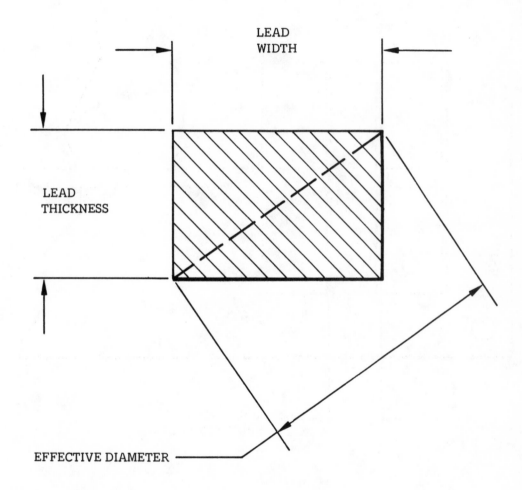

FIGURE 10-36
Effective IC Lead Diameter

Chapter

11

ARTMASTER TAPE-UP

ARTMASTER TAPE-UP AND PHOTO-REDUCTION

This chapter illustrates the mechanics of pc board pattern making. It shows how to create a master artwork interconnect pattern, and how to photo-reduce it to normal size.

When a printed circuit board has been "stuffed" with electronic components, its printed conductive traces will interconnect the components. These interconnect patterns are created larger-than-life, then reduced to normal size, a technique that improves accuracy.

Patterns of conductive traces begin as strips of black adhesive applied to sheets of clear plastic. The shadows of these patterns are photographically captured, reduced, and repeated on manufactured boards as conductive traces. These are the board's printed circuits.

Three different techniques for creating the master art patterns will be introduced, and their tradeoffs discussed. This chapter illustrates specifically what to do, how to do it, and what not to do while taping the art master. It reveals proven techniques for handling the tape and provides rules-of-thumb for pattern spacings and tolerances.

Although pc board designers are not normally responsible for the subsequent photographic processes, techniques and equipment are described so that designers can interface smoothly with the next step in pc board design.

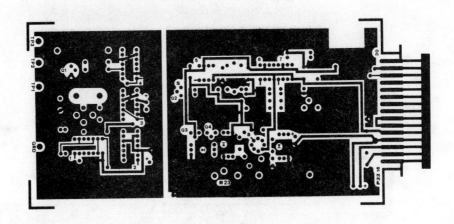

FIGURE 11-1
Artmaster

TAPE AND PADS

Black crepe tape, pads and patterns provide a material that is easy to work with and the traces can be routed either by the sweeping bend method or by the cut and bend method. The latter is preferred due to the fact that better proficiency in both time and appearance can be accomplished in a shorter learning time. With the black crepe method, a separate layer or she[et] must be prepared for each side of t[he] board. This can be accomplished by a tw[o] layer method (Figure 11-2) where the fi[rst] layer has all the pads plus the traces fr[om] one side of the board, and the second lay[er] has the same pads spotted with the trac[es] from the opposite side.

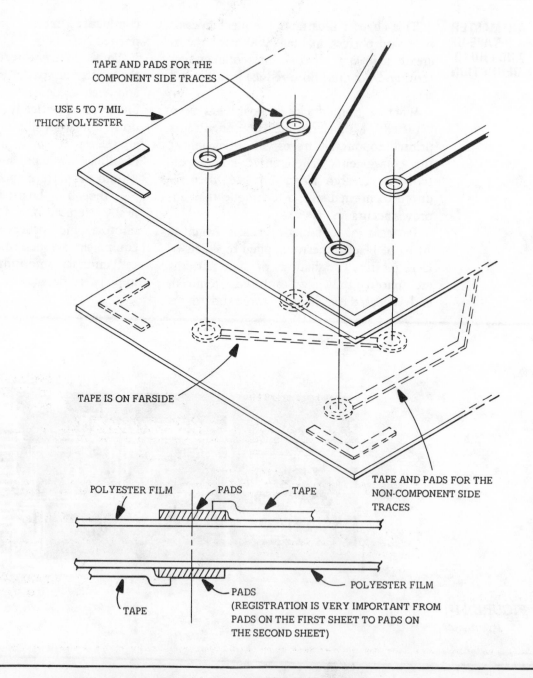

TAPE AND PADS FOR THE COMPONENT SIDE TRACES

USE 5 TO 7 MIL THICK POLYESTER

TAPE IS ON FARSIDE

TAPE AND PADS FOR THE NON-COMPONENT SIDE TRACES

POLYESTER FILM   PADS   TAPE

PADS

TAPE   POLYESTER FILM

(REGISTRATION IS VERY IMPORTANT FROM PADS ON THE FIRST SHEET TO PADS ON THE SECOND SHEET)

FIGURE 11-2
Two Layer
Method

ARTMASTER TAPE-UP

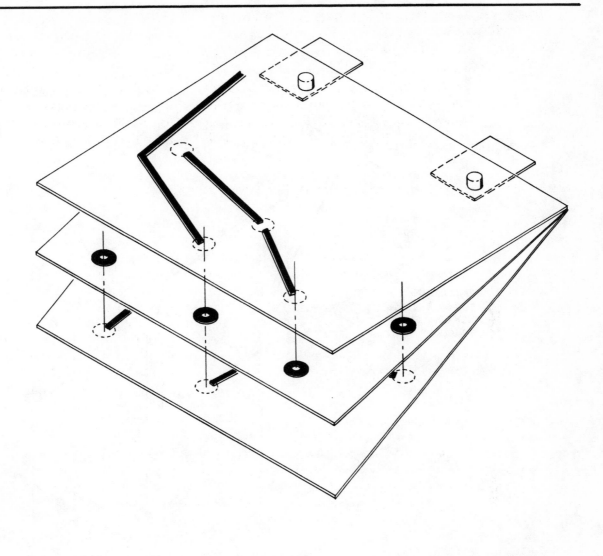

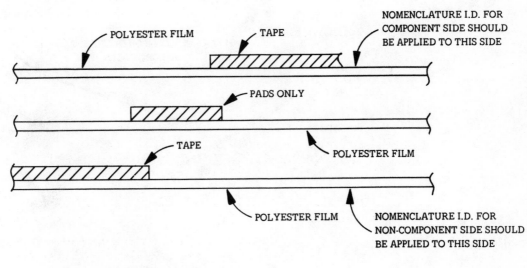

FIGURE 11-3
Three Layer
Method

POLYESTER FILM

TAPE

NOMENCLATURE I.D. FOR
COMPONENT SIDE SHOULD
BE APPLIED TO THIS SIDE

PADS ONLY

TAPE

POLYESTER FILM

POLYESTER FILM

NOMENCLATURE I.D. FOR
NON-COMPONENT SIDE SHOULD
BE APPLIED TO THIS SIDE

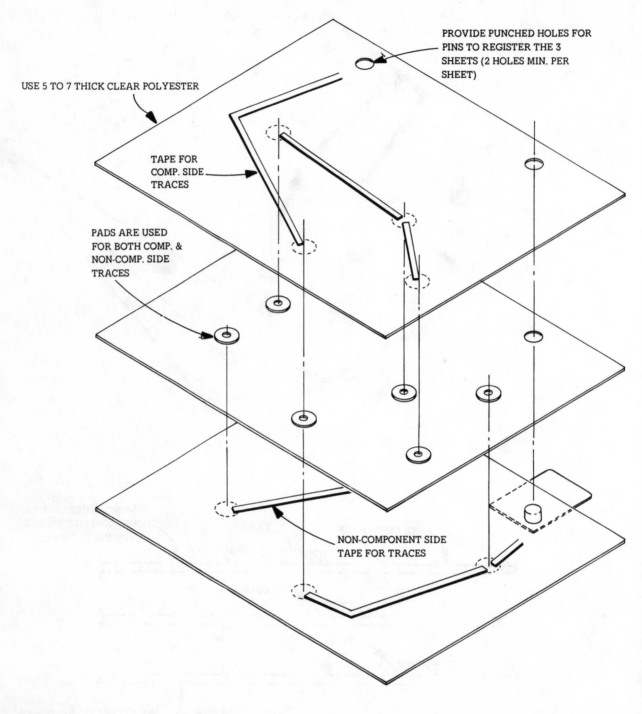

PROVIDE PUNCHED HOLES FOR PINS TO REGISTER THE 3 SHEETS (2 HOLES MIN. PER SHEET)

USE 5 TO 7 THICK CLEAR POLYESTER

TAPE FOR COMP. SIDE TRACES

PADS ARE USED FOR BOTH COMP. & NON-COMP. SIDE TRACES

NON-COMPONENT SIDE TAPE FOR TRACES

FIGURE 11-3a
Three Layer
Method

ARTMASTER TAPE-UP

The three-layer method shown in Figure 11-3 is essentially the same, except one layer is used for a "pad pattern," and then this is overlaid with a sheet for the non-component traces and a different sheet for the component traces. This method saves time by not having to "spot" the second set of pads opposite the first set, but most important, insures perfect registration of all pads from component side to non-component side because the same pad pattern is used for both sides with the respective trace pattern superimposed over it.

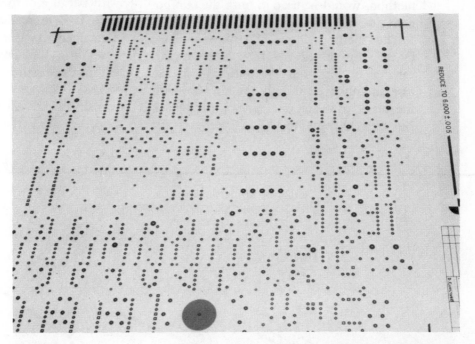

Pad Master

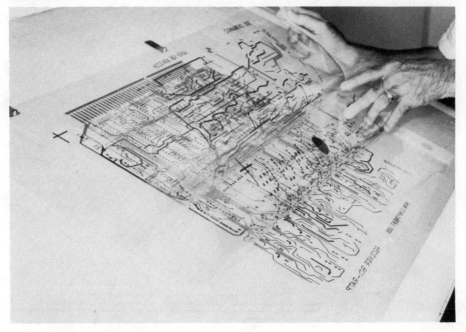

Pad master with component and Non-component sheets

ARTMASTER TAPE-UP

The red and blue material method illustrated in Figure 11-4 provides tapes in the same widths as black crepe except they can be taped on one sheet of polyester film with red on one side and blue on the other side. The pads would be black crepe placed only on one side and, as in the three layer method, would be used in both the component side and non-component side.

This method allows the photographer, by the use of red and blue filters on his camera, to filter out the blue tape in one case and photograph only the red tape with the black pads for one side of the board. He then changes the filters, repeats the process and photographs only the blue tape with the black pads for the opposite side of the board. The tape-up does not have to be flipped over or moved during this entire process. The end result is absolute registration for two-sided P.C. Boards.

Red and blue tape is not as easy to work with because of its rigid polyvinyl composition. It is more difficult to cut and takes more pain to ensure that it will stick to the base film. Red and blue cannot be bent without distortion of the traces; therefore it has to be cut all the way through and repositioned each time you change direction, or, using the cut and bend method

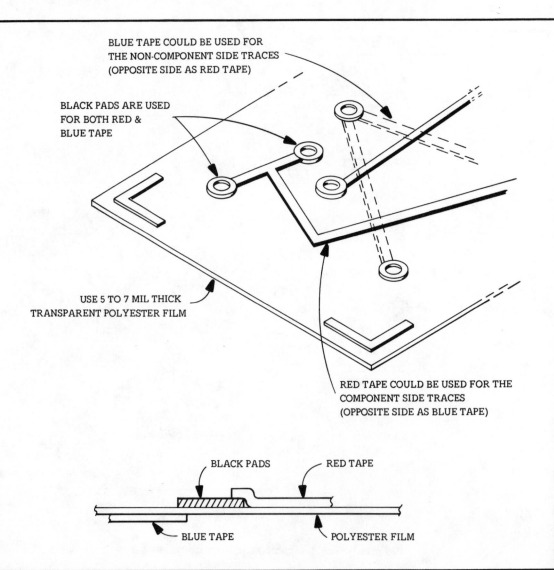

BLUE TAPE COULD BE USED FOR THE NON-COMPONENT SIDE TRACES (OPPOSITE SIDE AS RED TAPE)

BLACK PADS ARE USED FOR BOTH RED & BLUE TAPE

USE 5 TO 7 MIL THICK TRANSPARENT POLYESTER FILM

RED TAPE COULD BE USED FOR THE COMPONENT SIDE TRACES (OPPOSITE SIDE AS BLUE TAPE)

BLACK PADS   RED TAPE

BLUE TAPE   POLYESTER FILM

FIGURE 11-4
One Sheet Red &
Blue Method

cut the tape approximately 90 percent through and then bend the remainder to the required degree necessary and burnish down the edges.

Regardless of taping methods or type of material used, the physical act of taping must conform to some basic standards.

Conductor angles should be made at preferred angles of 45 or 90 degrees. Although the traces could be directed at almost any angle, it is important that all traces running parallel be at the same angle for uniformity. The minimum angle that any trace should be placed at is 90 degrees. Angles less than 90 degrees provide a situation during the manufacturing process which could allow etching solution to build up on the inside angle and etch away excess material as illustrated in Figure 11-5(B).

CONDUCTOR ROUTING

The same problems discussed above also apply to trace routing. If the junction of traces coming off a pad form a pocket, the same over-etching problem can occur as shown in Figure 11-6B.

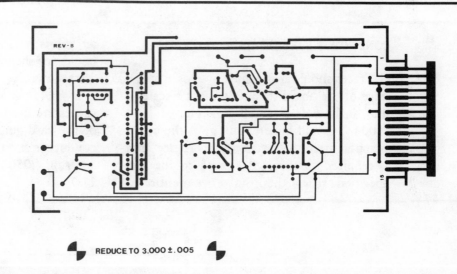

REDUCE TO 3.000 ± .005

Typical Tape-Up

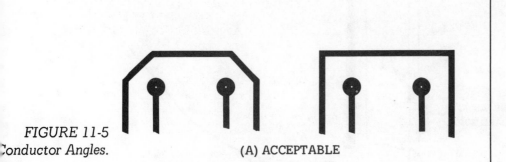

FIGURE 11-5
Conductor Angles.

(A) ACCEPTABLE

(B) NOT ACCEPTABLE

✷ LESS THAN 90

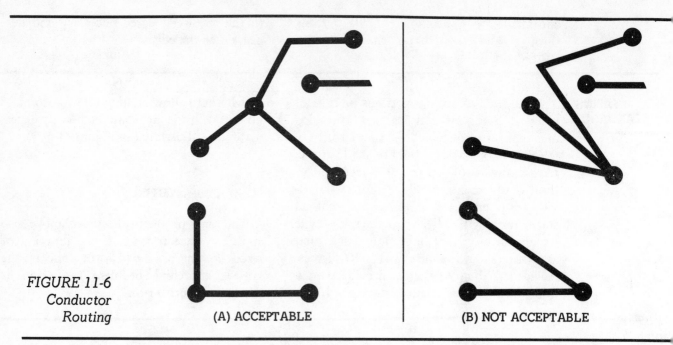

FIGURE 11-6
Conductor
Routing

(A) ACCEPTABLE   (B) NOT ACCEPTABLE

CONDUCTOR SPACING TO BOARD EDGE

Traces should not come any closer to the edge of the board than .050-in. at a scale of 1x1 inch and, if at all possible, at least .100-in. should be left. This gives the board manufacturer the necessary tolerance that is required to shear the board to the size specified. If less than the above-mentioned air gap is left, a portion or all of the tra could get sheared off or shorted out at t. next assembly.

If metal or electrically conductive ca guides are used, then the gap from the ed of the card guide to the edge of the neare conductor should be the standard requir air gap (.050-in. or greater); see Figu 11-7.

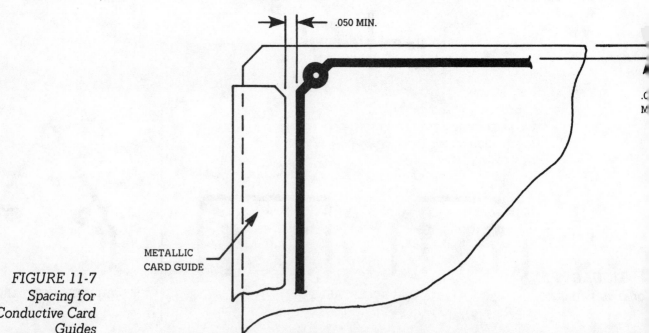

.050 MIN.

METALLIC
CARD GUIDE

FIGURE 11-7
Spacing for
Conductive Card
Guides

TAPING DON'TS

There should be no overlapping of tape or cutting of pads. Tape has a natural tendency to "creep" or return back to its original shape if it is bent or stretched at all. Even under the most careful conditions, this can be a problem so care must be taken not to create even more difficult situations where tape could distort and cause electrical and manufacturing problems. See Figure 11-8.

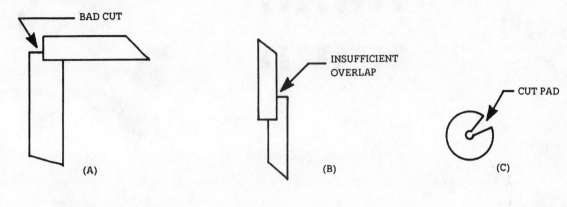

FIGURE 11-8
Taping Problems

APPLYING MULTI-PAD STIKON PATTERNS TO THE ARTWORK

To remove a multi-pad stikon pattern (See Figure 11-9) from the tape in order to prepare it for positioning on a grid pattern, hold the backing paper in one hand, grasping the top of the tape between the index and second fingers. In the other hand, hold the knife so that the index finger rests on the flat side of the blade. This hand will face downward with the thumb resting against the inner side of the knife handle and the remaining fingers curled under the knife. The pointed edge of the knife should be positioned toward the inside. (See Figure 11-10.) With a finger supporting the back of the tape, slide the tip of the knife under one edge of the stikon. While holding the pattern against the knife blade with a finger, gently peel the stikon away from the backing. To enable easy positioning of the stikon, and to prevent repeated positioning, replace the stikon on the backing so that one row of pads hangs over the edge of the backing. (See Figures 11-11 and 11-12.) Position this single row of pads over the grid so that an entire section of lines is centered in each pad hole. (See Figure 11-13.)

Remove the backing from the rest of the stikon and lay it down on the grid. Check to be sure it is accurate. Do not press the pattern firmly into place in case repositioning is necessary. The pattern can be easily lifted and placed in the correct position by sliding the knife blade under the stikon and gently peeling it toward you. Realign as described for the initial positioning. When correctly placed, apply pressure firmly over entire stikon to affix it permanently to the surface.

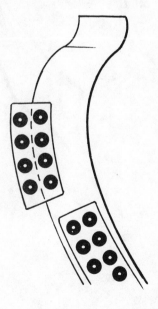

FIGURE 11-9

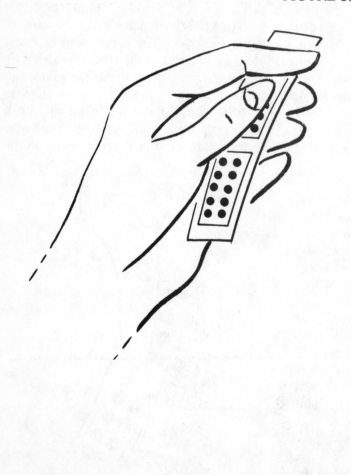

FIGURE 11-10

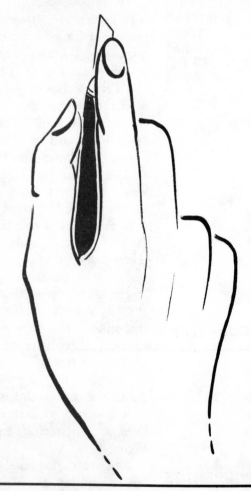

　　　　　　　　ARTMASTER TAPE-UP

FIGURE 11-11

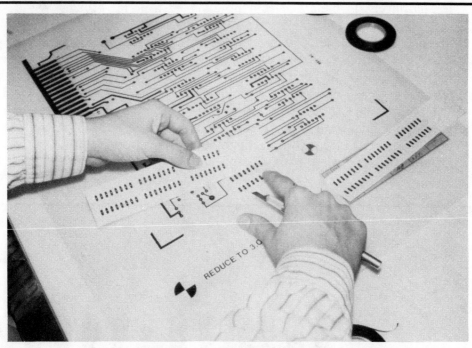

FIGURE 11-12

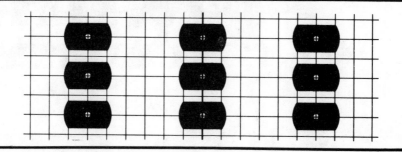

FIGURE 11-13

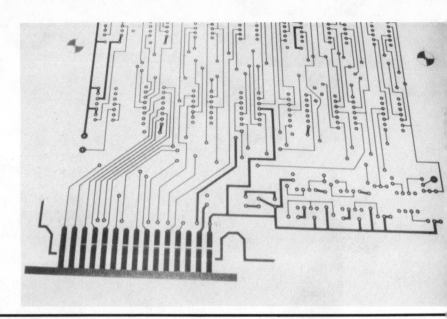

To remove a connector strip from the backing, hold the strip in one hand. Grasp the top of the strip between the index and second fingers. In the other hand, hold the knife so that the index finger rests on the flat side of the blade. This hand will face downward with the thumb resting against the inner side of the knife handle and the remaining fingers curled under the knife. The pointed edge should be positioned toward the inside. With a finger supporting the back of the tape, slide the tip of the knife under one edge of the connector strip. While holding the connector strip against the knife blade with your index finger, gently peel the connector strip away from the backing.

To facilitate fast and easy positioning replace the connector strip on the backing so that just the tips of the connector extend over the edge of the backing. (See Figures 11-14 and 11-15.) Position these tips on the artwork precisely in the desired locations. Press into place. Remove the backing from the remaining connector strip and press it firmly into place. (See Figure 11-16.

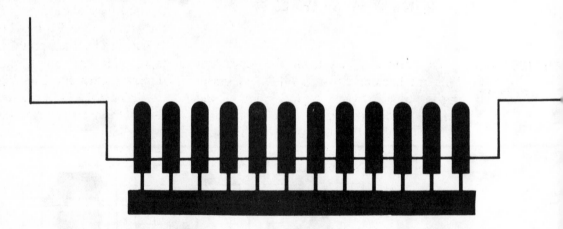

FIGURE 11-14

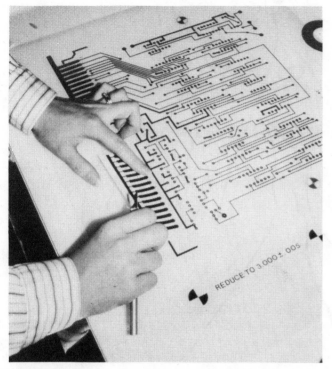

FIGURE 11-15

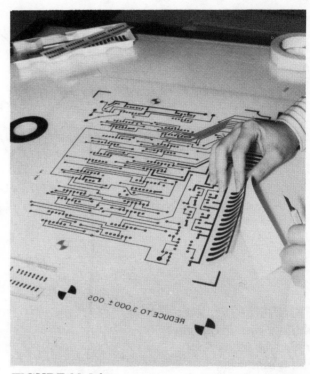

FIGURE 11-16

To remove a pad from the tape, first hold the backing paper in one hand, grasping the tape between the index and second finger. In the other hand, hold the knife so that the index finger rests on the flat side of the blade. This hand should face downward with the thumb resting against the inner side of the knife handle and the remaining fingers curled under the knife. The pointed edge should be positioned toward the inside. (See Figures 11-17 and 11-18.)

With a finger supporting the back of the tape, slide the point of the knife very slightly under the pad. While holding the pad against the knife blade with your index finger, lift the pad up. Use the knife blade as a holding tool and visually align the pad center with its grid intersection. (See Figure 11-19.)

ARTMASTER TAPE-UP

195

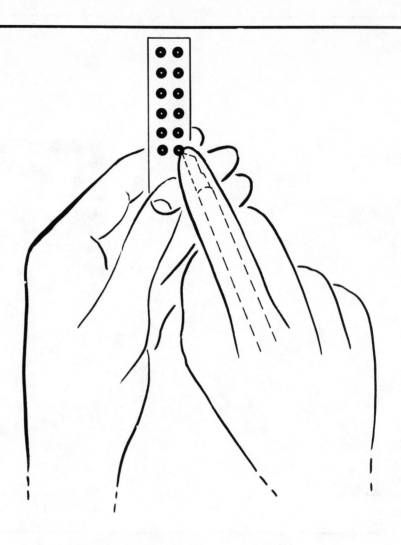

FIGURE 11-17
*Knife shown in outline
under the index finger.
Pad is sandwiched
between knife and finger
when blade slides under
pad edge.*

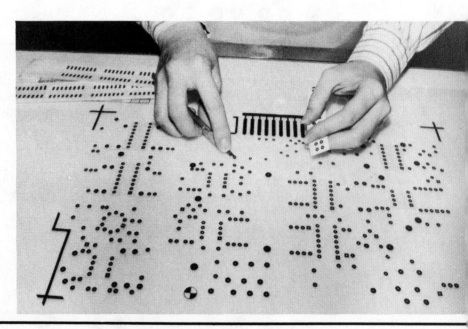

ARTMASTER TAPE-UP

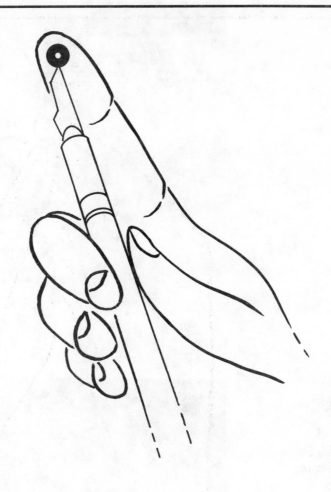

FIGURE 11-18

...sandwiched between ...er and knife to lift ...om the backing—lay ...in position and ...s down with index ...er from other hand

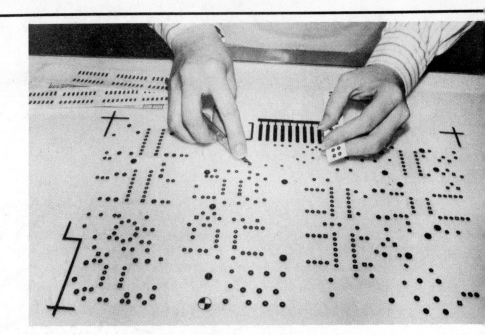

FIGURE 11-19
Use the knife
blade as a holding
tool and visually
align the pad
center on the grid.

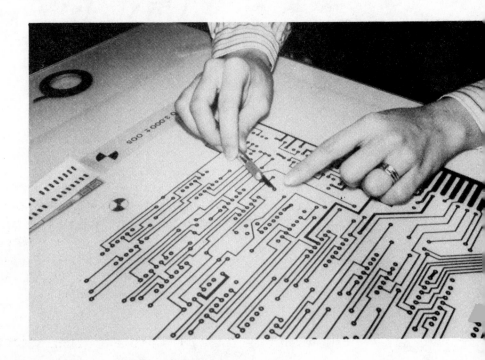

TAPING EXERCISE

When laying a strip of tape, it is important to hold the tape roll correctly to increase taping speed and avoid awkwardness. In your writing hand, hold the knife as you would a pencil. (See Figure 11-20.) In the other hand, hold the roll of tape so that the index finger is through the hole and the thumb is resting on the top of the roll, as a brake. (See Figure 11-20.) The tape should be positioned to pull off the roll clockwise. (An easy way to locate the end of the tape is to lay it flat on the light table where the end will be more visible along the edge of the roll. (See Figure 11-22.)

See taping exercise sheet. (Figure 11-21) Remove it from book (cut along dashed line) or make a photocopy and place a sheet of mylar over it so you can practice taping. Use .100-inch and .050-inch tape. *Taping straight lines:* With the index finger used as an axle, draw out the tape from the roll with the thumb and index finger of the hand holding the knife until the desired length is obtained. Use your thumb to control the unwinding. (See Figure 11-23.)

To help obtain the correct length of tape without pulling off too much—as you pull out the tape, keep it just above the desired line of placement and pull off an inch or two extra. (See Figures 11-24 and 11-25.)

Lower the loose end of the tape to a point slightly past the end of the desired line of placement while centering the tape vertically. Press to tack down. (See Figure 11-26.) While holding this end, stretch the tape slightly to keep it straight, being careful not to break or distort it. Lower the opposite end of the tape while centering it on the desired line. Burnish the tape after the trace is established. (See Figure 11-26.)

To trim the tape at the ends, the "fixed-cut" should be used to avoid cutting the mylar or other stikons. To do this, hold the knife edge firmly with the blade in a straight line across the width of the tape. Then pull the tape up and along the blade edge so as to insure a clean cut. (See Figure 11-27.) First cut the end connected to the roll. (See Figure 11-28.) Then, with the blade edge facing the mylar, lift the other end of the tape, grasp it with the other hand, position the knife where the trace needs to be cut, and pull the tape up to cut it. (See Figure 11-29.) At no point during the taping process should you put down either the roll of tape or the knife. This is to reduce unnecessary movement, and enable the taping process to proceed quickly and efficiently.

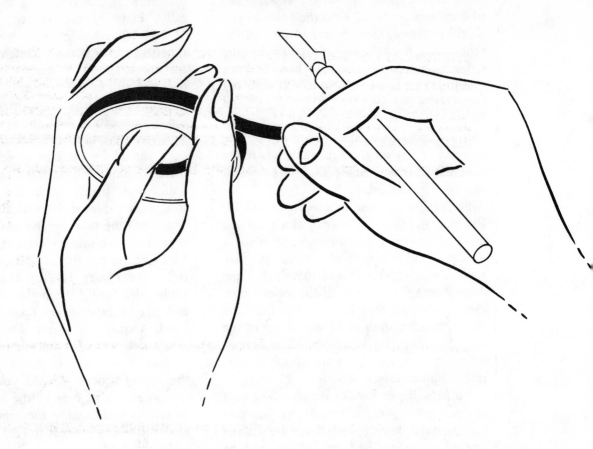

FIGURE 11-20

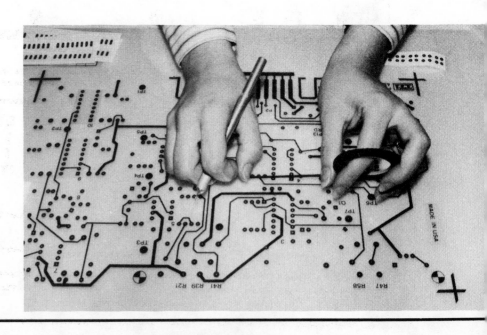

ARTMASTER TAPE-UP

TAPING PRACTICE

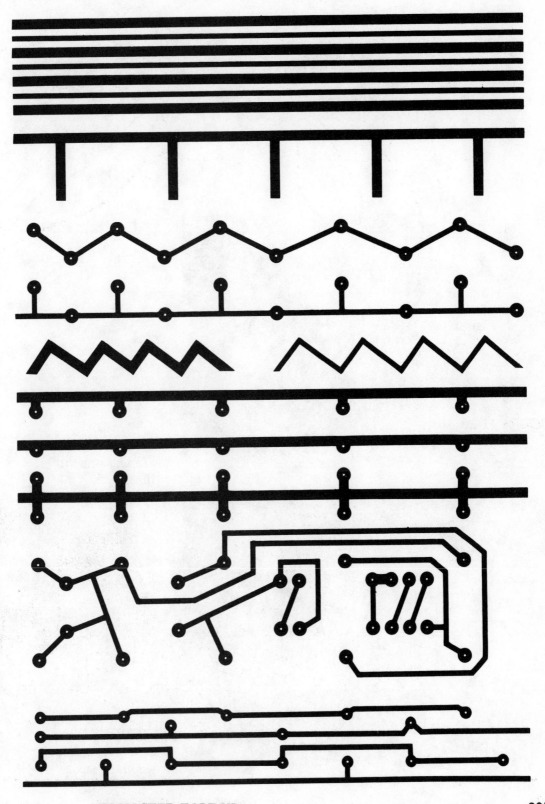

FIGURE 11-21

ARTMASTER TAPE-UP

FIGURE 11-22
Locating end
of tape.

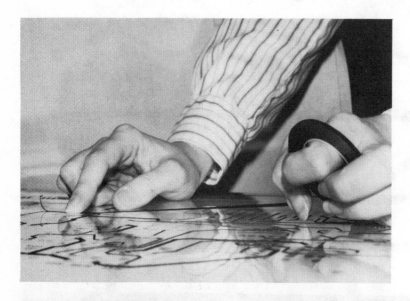

FIGURE 11-23
Taping straight
lines.

FIGURE 11-24
Obtaining correct
length of tape.

ARTMASTER TAPE-UP

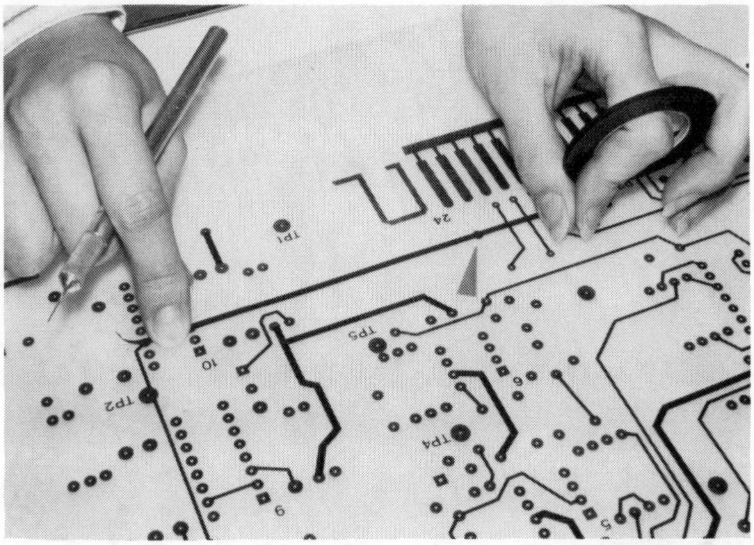

FIGURE 11-25
ut-off point
own by arrow
enter).

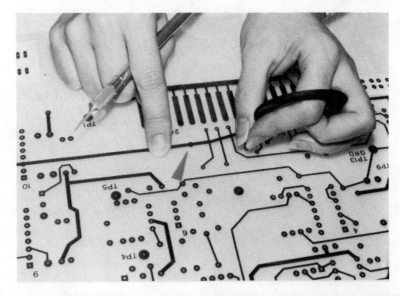

FIGURE 11-26
urnish to tack
own.

FIGURE 11-27
ut off.

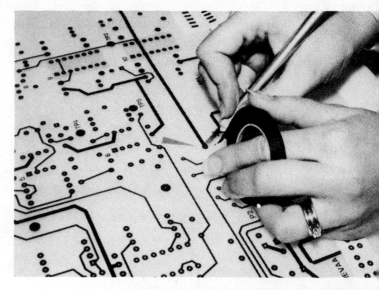

FIGURE 11-28
Cut the end
connected to
the roll.

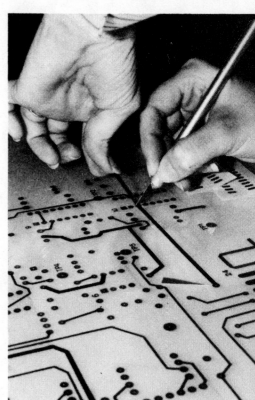

FIGURE 11-29
Final cut.

ARTMASTER TAPE-UP

TAPING TRACE LINES

Two common widths of tape are .100 and .050 inch. The thinner tape requires gentle, firm handling.

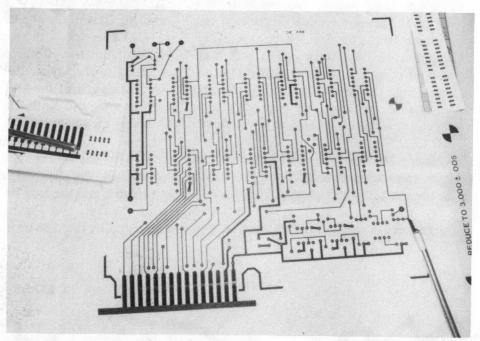

Examples of .050 and .100 tape widths

Tape requires gentle but firm handling

ARTMASTER TAPE-UP

Changing direction of tape line: A more practical alternative to the free-flow curve method of taping (See Figure 11-31) is the angle-cornering method. (See Figure 11-30A and 11-32.) It reduces stress on the tape that is caused by the memory of the tape. When laid in a curve, the tape wants to pull itself back into a straight position and does not adhere firmly to the mylar. Cutting the tape nearly through, toward the edge of the new direction, and then turning the tape in the new direction relieves stress, saves space, and is faster. (See Figure 11-33) In the free-flow style, turns are not as tight and 90° curves are impossible. (See Figure 11-30B)

To turn the tape, first lay the initial trace direction and cut the trace where you

wish the outside edge of the tape to foll in the new direction. Cut nearly all the w through the width of the tape, perpendi lar to the edge, by pushing the bla straight down on the tape. (See Figu 11-34) This deep cut will prevent the ou edge of the turn from rolling up and r sticking to the mylar. Lift the tape comi off the roll with the knife edge facing t mylar and also with the hand which holding the tape roll, and then turn t tape in its direction. (See Figure 11-35 a 36) After laying the tape in its new dir tion, and you are certain it is position correctly, use the heel of the knife to b nish the tape. (See Figure 11-37) If it is r straight, lift the tape up and reposition it

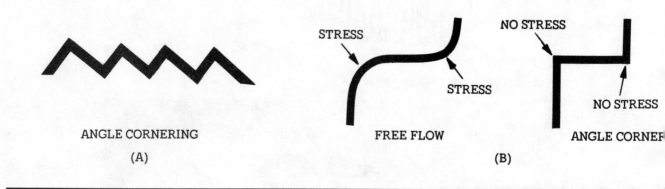

ANGLE CORNERING

(A)

FREE FLOW

ANGLE CORNEF

(B)

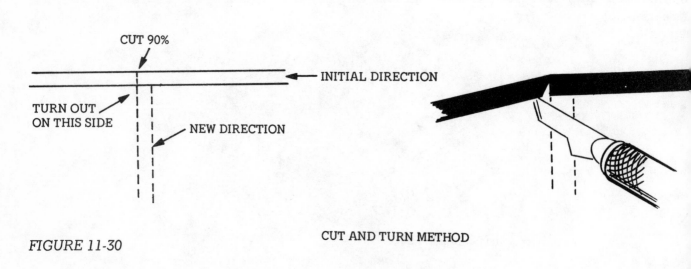

CUT 90%

INITIAL DIRECTION

TURN OUT ON THIS SIDE

NEW DIRECTION

FIGURE 11-30

CUT AND TURN METHOD

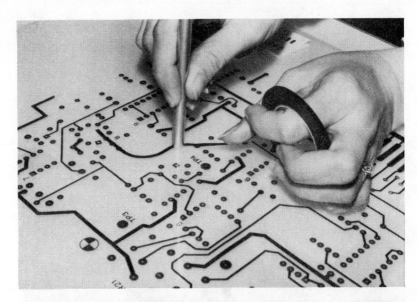

FIGURE 11-31
ree-flow curve
ping method.

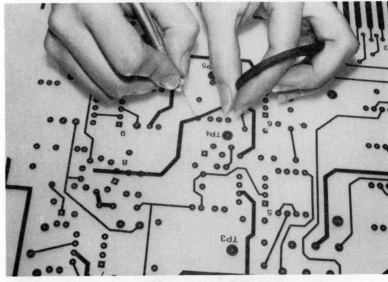

FIGURE 11-32
ngle-cornering
ethod.

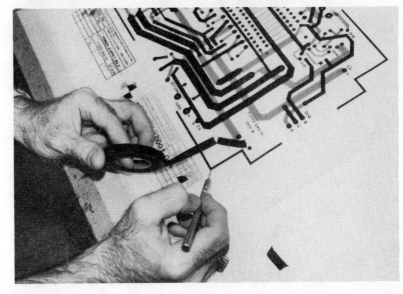

FIGURE 11-33
ut and turn.

ARTMASTER TAPE-UP

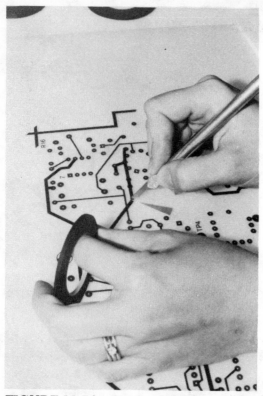

FIGURE 11-34   Cutting 90%

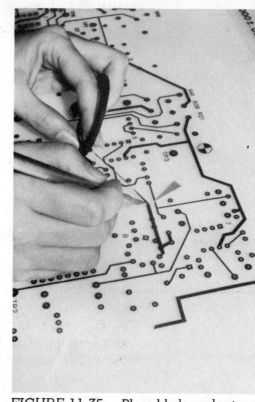

FIGURE 11-35   Place blade under tape

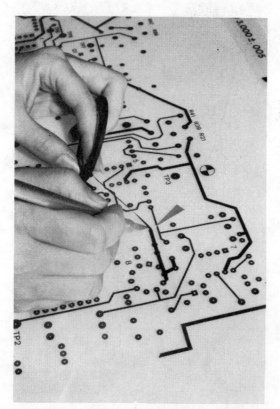

FIGURE 11-36   Turn to desired direction

FIGURE 11-37   Burnish corner

ARTMASTER TAPE-UP

PLACING A TRACE PERPENDICULAR TO ONE PREVIOUSLY LAID:

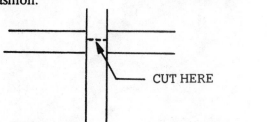

POOR CUT

Lay the new trace perpendicular to the previous trace in a cross-like fashion.

CUT HERE

Cut the new trace where it meets the center of the previous trace, as illustrated by the dashed line above. Use the "fixed cut" method: Hold the knife blade firmly in a fixed position with the blade in a straight line across the tape. (See Figure 11-38.) With the other hand, pull the tape up and along the blade at an angle to insure a clean cut. (See Figure 11-39.)

FIGURE 11-38

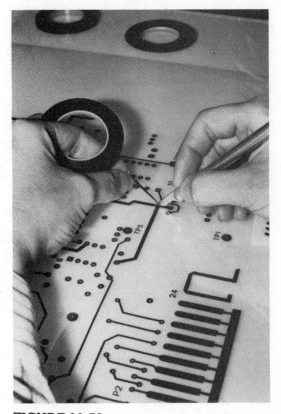

FIGURE 11-39

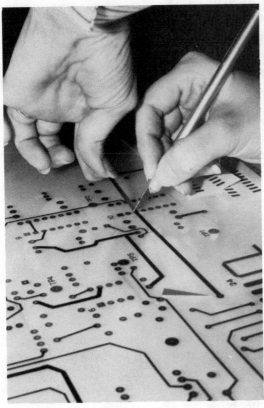

Don't use a pointed-type blade #11.

Use a flat-type blade #16. This will

help prevent cutting through mylar and pads and artwork.

While placing pads, keep pad strip an knife in your hands at all times. Do not la pads or knife down at the same time.

Burnish all pads.

"While Taping" Keep the roll of tape and knife in your hands. "Don't" lay the tape or knife down while taping.

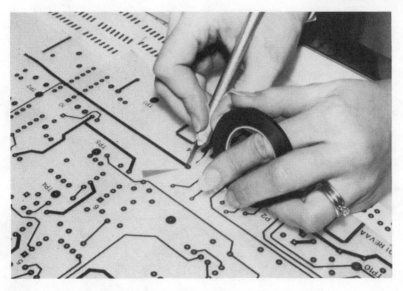

Don't cut through pads by pushing too hard.

Burnish all cut corners.

Don't over-stress tape on free flow bends.

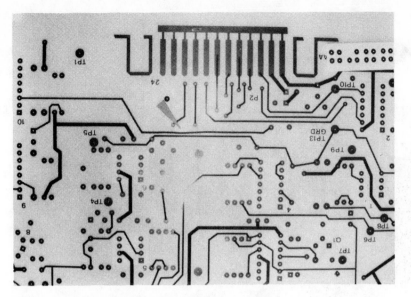

Tape must be straight, and parallel to other
tape. Don't allow tape to wander.

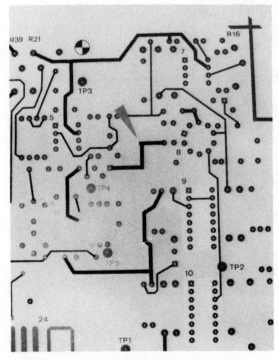

Cut corners clean. Don't overlap.

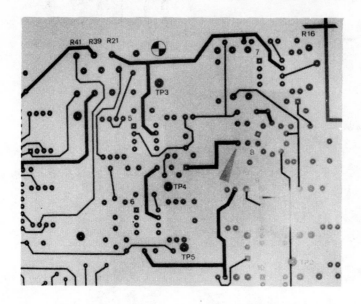

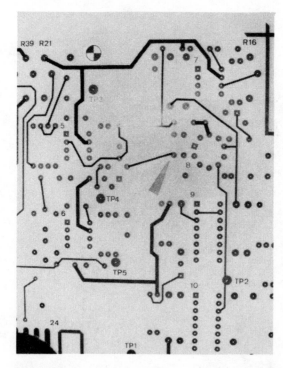

Tape to center of pads (top). Don't tape to one side (bottom).

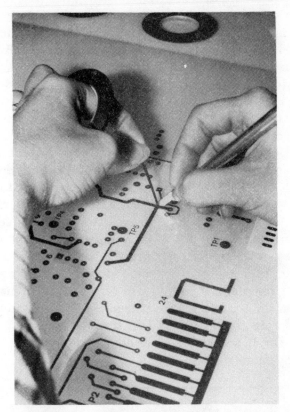

Pull tape away from knife. Don't pull knife
through tape and cut mylar.

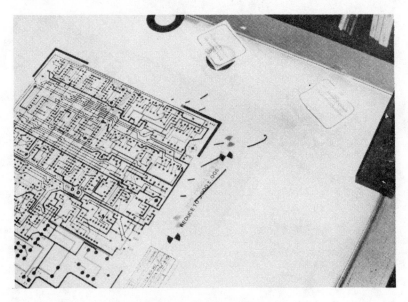

Tape-up should be clean and free of rem-
nants. See Figure 11-31.

Some applications require all reference designations of the electrical components either be etched or silk-screened on the P.C. board for easy identification. This aids technicians and field service repairmen to repair or trouble-shoot boards without the use of an assembly diagram. Boards that have a high failure rate or have to be repaired without access to documentation rely on this information. In cases where such rigid identification is not necessary, it is still important to identify such components as connectors, test points, switches, or any component that will require access for calibration or trouble-shooting. If a particular board requires that jumpers be installed to reflect different models or functions, then they should be marked on the board also.

Integrated circuit packages should be identified on the board. The IC's are not generally oriented in any numerical se-

quence on the schematic. Therefore, order to relate a reference designation fr the schematic to the P.C. board, the ir grated circuit (IC) mechanical pack must be identified by a reference desig tion number on the board.

It is helpful to identify some of contacts on a connector or any ot cluster of pins that are pre-assigned a nu ber arrangement. Generally, the first last pins are enough to establish this quence; in other cases, the first or last in each row might be required. If a parti lar board uses jumpers to distinguish e trical difference between models, it i good idea to identify these.

In general, keep in mind that little eff is required to clarify a problem at the sign stage, whereas downstream a solut could be costly.

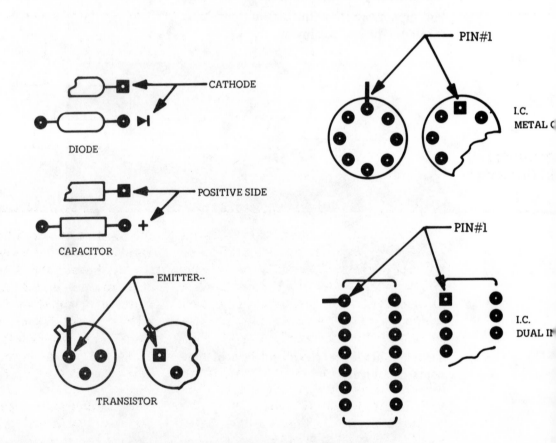

FIGURE 11-40

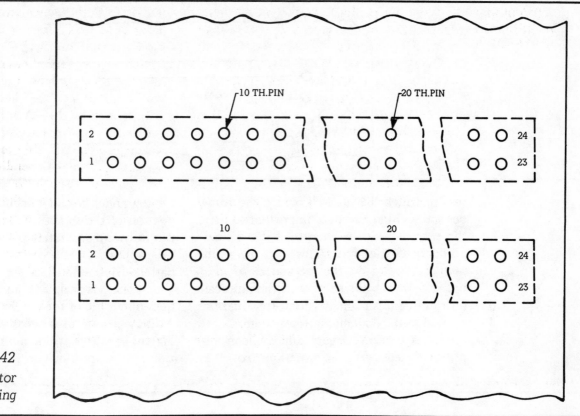

FIGURE 11-42
Connector
Indexing

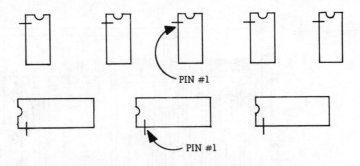

FIGURE 11-43
IC Orientation

After the board is taped and ready for the photography step, the final documentation information must be added to the tape-up.

The revision letter of the board must be identified on the artmaster somewhere within the board perimeter. This allows it to be etched on the board, to avoid confusion with previous or later changes. This is generally done as in Figure 11-44 on the non-component side of the board, along with the P.C. board part number.

The assembly number of the board should be identified in the same manner on the component side. This gives quick reference to anyone in finding the proper assembly drawing and any other documentation involved in the manufacturing, procurement or sale of the board. No assembly revision letter should be etched on the board; this information should be added using the actual revision the assembly drawing is at the time the board is assembled.

The master pattern number refers to the artmaster itself and will document the number and revision letter of the finished

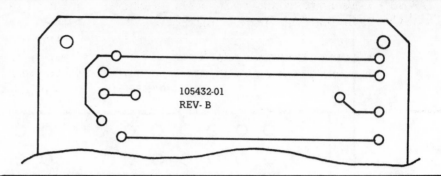

FIGURE 11-44
Revision Markings

105432-01
REV- B

tool to which the board is made. The number and revision letter are important so that the artmaster can be identified if the final product should fail. This identification should be outside the perimeter of the board as it will not be part of the physical board once it is manufactured. It should be entered in the documentation system.

The reduction targets and dimensions provide necessary information to the photographer to aid him in the proper r[...] duction of the artmaster. Targets should b[...] carefully and accurately placed so that the[...] will appear on all copies of the mast[...] pattern. Reduction marking should exten[...] along one vertical side and/or along o[...] horizontal side of the tape-up.

Examples of an artwork master contai[...] ing the master pattern number and redu[...] tion targets are shown in Figure 11-45.

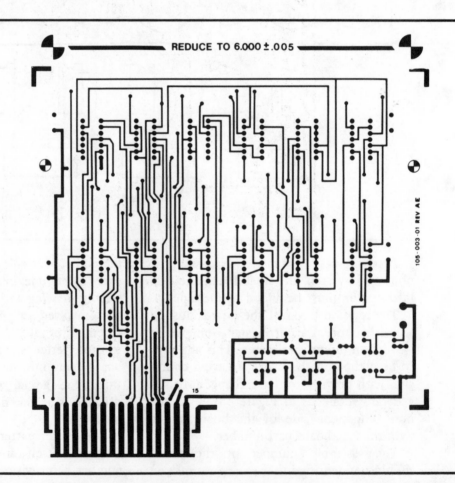

REDUCE TO 6.000 ±.005

105-003-01 REV AE

FIGURE 11-45
Artwork Master
Showing Reduction
Targets

TAPING GUIDELINES

1. Minimize handling of mylar and tapes.
2. Do not fold, crease or tightly roll artwork.
3. Burnish tape when complete.
4. Run tapes to center of annular ring of pad.
5. Do not cut tapes or pads through.
6. Do not pull tape *hard* or it will creep.
7. Place pads and tape on grid coordinates or coincident with lines.
8. Trim all corners.
9. Be neat.

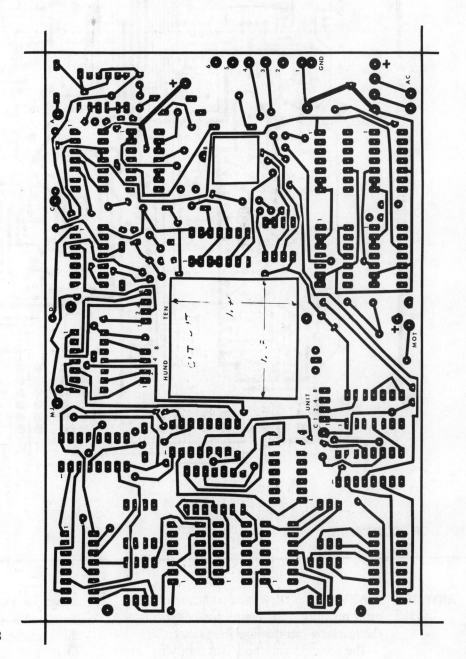

EXAMPLE OF POOR TAPING

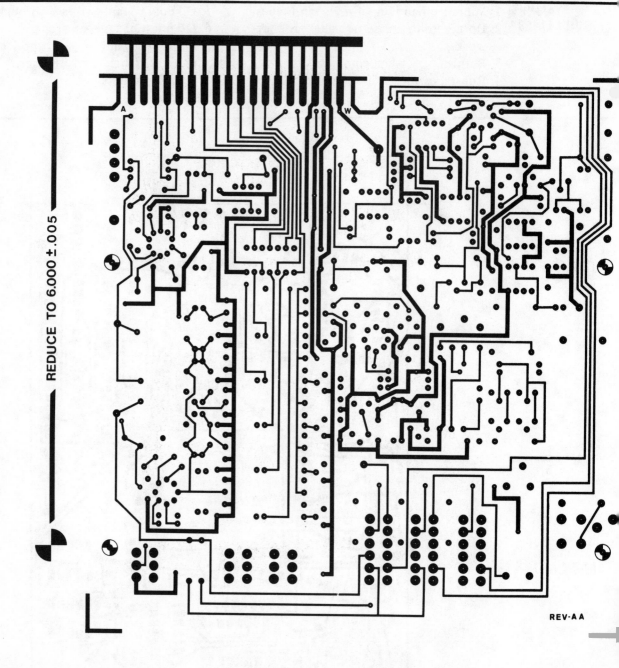

REDUCE TO 6.000 ±.005

REV-A A

EXAMPLE OF GOOD TAPING

ARTWORK COPIES

NEVER run the tape-up through a blueprint machine. The lamps and rollers inside the machine can damage the tape-up.

The correct method for copying art-master tape-ups is to use a (Plate Maker) shown in Figures 11-46. 11-47 and 11- This technique is used because the tape- is flat at all times and under no stress.

FIGURE 11-46

Glass door is raised and artwork is positioned.

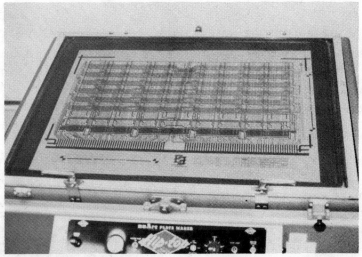

FIGURE 11-47

Glass door is lowered and the tape-up is placed under vacuum.

FIGURE 11-48

Top is turned over and tape-up is exposed.

The completed PC artwork is the master tool, affecting all subsequent stages through production of the printed circuit board. Every way in which tolerance build-up can be reduced or eliminated is invaluable.

The proper handling of all of the materials which are used to create the master artwork will produce the optimum results.

Only drafting film and glass have the dimensional stability required for the development of printed circuit artworks. This includes precision grids and scales, drafting films and photographic and diazo films.

The following procedures are recommended to safeguard all drafting film products.

- Store cartons in an area where the temperature and humidity remain relatively constant.
- DO NOT expose to temperatures in excess of 120°F (40°C) or below 33°F (1°C).
- Films should be stored and transported flat in a closed container.

All sensitized and unsensitized drafting film products should be conditioned to the environment in which they will be used.

- Move the materials into the work area far enough ahead of time (preferably 24 hours) to allow for complete equillibrium.
- Open packages of unsensitized materials and separate individual sheets of film. (Rolls should also be opened and cut to appropriate sizes.)
- Materials which are used together should come from the same package or roll and should be used in the same orientation. This can be accomplished by notching the corner when cutting or removing from the package.
- DO NOT expose your grid to excessive sunlight as some fading or discoloring may occur.

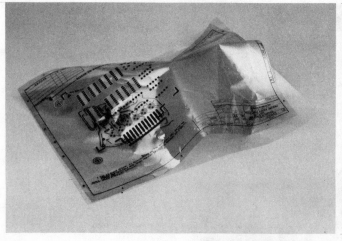

PREVENT THIS   USE THIS

FIGURE 11-49

CAMERA PC artwork is photographically reduced using a graphic arts process camera. This type of camera is explicitly designed for copying line or half-tone material. It must be capable of repeatedly producing the same photo reduction. The camera must be dimensionally stable and have distortion free optics. (Figures 11-50, 11-51, 11-52, 11-53 and 11-54).

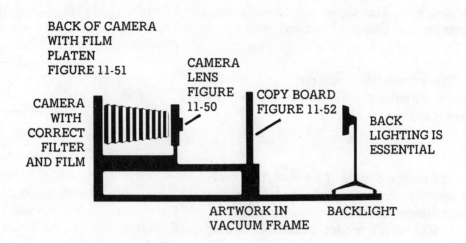

BACK OF CAMERA WITH FILM PLATEN FIGURE 11-51

CAMERA LENS FIGURE 11-50

COPY BOARD FIGURE 11-52

CAMERA WITH CORRECT FILTER AND FILM

BACK LIGHTING IS ESSENTIAL

ARTWORK IN VACUUM FRAME   BACKLIGHT

TECHNIQUES Back light the copy board to avoid fuzzy edges, particularly on taped-up master artwork. Back lighting provides contrast of 1,000:1, compared with approximately 50:1 for front lighting.

A green light should be used to achieve the optimum results *except* when the Red and Blue separation method is used.

Always specify the side of the film on which you want the emulsion (image). The film should always have the emulsion in direct contact with the emulsion of that to which it is being contacted. Normally, negatives are right reading (R/R), emulsion down and positives, right reading, emulsion up.

Applying the images to the printed circuit board in register can be accomplished by many methods (see Registration Techniques). One of the best uses identically prepunched boards and master patterns. The working copies of the master patterns are made using the PC board indexing hole system to expose pre-punched photo film.

Another method of applying the image is to make the master pattern visually transparent by "bleaching the photo." This allows the operator to actually see through the circuit pattern, and shift the photo around to obtain the best registration. Diazo coated film is available which is ideally suited to this method.

PHOTO SEPAR-ATION OF RED AND BLUE ARTWORK Properly prepared Red and Blue master artwork will produce two negatives in perfect register by following these instructions:

• Camera lens must be color corrected for lateral color aberration. Backlighting from a constant white light source is essential for color separation.

- Proper filters are vital—use only Kodak's Wratten Gelatin filters, PM47 Blue and PM29 Red. These are uncemented gelatin film filters and are relatively inexpensive.

NOTE: The blue filter is often not necessary because this film is not sensitive to blue light. This allows shorter exposure times and less chance of over-exposure.

(Environmental, material, and equipment variations may necessitate adjustments to the data listed.)

MASTER PATTERN INSPECTION CHECKLIST

The master pattern is a production tool. It should be inspected and controlled in the same manner as other precision tooling.

1. **SIZE**—Each master pattern should be measured at the photographic reduction marks, and at least one other dimension at right angles to the first measurement to insure proper reduction and size.
2. **REGISTRATION**—All patterns required to produce a printed circuit board must be inspected together to insure that registration as stated on the master drawing has been achieved.
3. **SHARPNESS**—Image sharpness and resolution are necessary to prevent ragged conductor edge definition on the etched printed circuit board. Sample conductor line width and spacing measurements must be made to verify proper exposure and development.
4. **DENSITY**—The image on the master pattern must be opaque enough to insure that proper exposure of the photosensitive coating can be made.
5. **WORKMANSHIP**—All master patterns should be inspected for pin-holes, scratches, fingerprints and general cleanliness. All touch-up work should be made on the non-emulsion side of the film.

TO DROP RED AND HOLD BLUE

Film: Kodalith Estar Base Panchromatic #2568 (check exposure data packed with film). If required, .007" (0.18mm) base pan film is available from DuPont. For test trial: camera setting same size.
Filter: #29 (Kodak-Wratten)
Exposure: f22 for 7 seconds
Develop: 2½ minutes at 68°F (20°C) Kodak Kodalith Super Developer (powder), develop with continuous agitation.

TO DROP BLUE AND HOLD RED

Film: Kodalith Type 3, Estar Base. For test trial: camera setting same size.
Filter: # 47 (Kodak-Wratten)*
Exposure: f16 for 17 seconds
Develop: 2¾ minutes at 68°F (20°C) Kodak Kodalith Super Developer (powder), develop with continuous agitation.
Light Source: 2-1,000 watt pulsed xenon lamps.

FIGURE 11-50
Camera lens

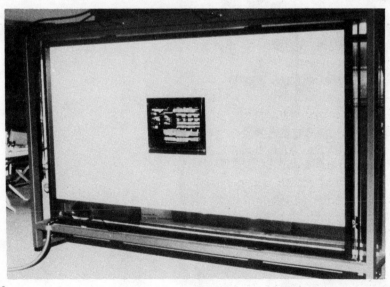

FIGURE 11-51
Camera copy board

FIGURE 11-52
Copy board with fixed registration pins for location of artwork

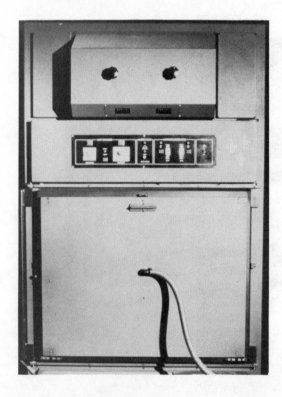

FIGURE 11-53
Back of camera

ARTMASTER TAPE-UP

FIGURE 11-54
Back of camera
and film platen
(new film to be
exposed)

FIGURE 11-55
Small (24" film)
processer

ARTMASTER TAPE-UP

FIGURE 11-56
Large (60" film)
Processer

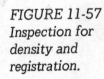

FIGURE 11-57
Inspection for
density and
registration.

ARTMASTER TAPE-UP

FIGURE 11-58
Inspection for
sharpness and
density.

FIGURE 11-59
Inspection for
workmanship:
pinholes, scratches,
fingerprints, etc.

Chapter

12

GROUND PLANES

The multi-point ground system is used primarily for shielding or high frequency type digital circuits. This is where we have IC's or circuits terminating to a massive area of copper referred to as a plane. Figure 12-1 will demonstrate this technique.

Ground planes are areas on the P.C. board that have not been consumed by traces or pads and are left with copper that is tied to the ground circuit (See Figure 12-2). The plane is generally on the component side, especially for flow soldering, and is isolated from the other traces and pads. Planes on the component side of the P.C. board are typically a continuous conductive area. Planes on the non-component side of the P.C. board wider than .05-in. should be broken up into a striped or checkered pattern (see Figure 12-3). This will help prevent blistering and warping during the flow soldering operation.

Adequate clearance should be provided around nonfunctional terminating areas in ground planes on external board layers. On

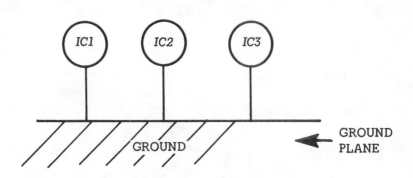

FIGURE 12-1
Multipoint Ground

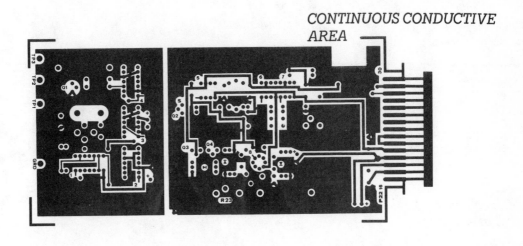

FIGURE 12-2
Ground Plane

CONTINUOUS CONDUCTIVE AREA

internal layers of multilayer boards, use of a nonfunctional terminal area is not necessary if a diameter is left clear around the hole. See Figure 12-4.

When a hole terminates in a ground plane and electrical continuity is required, a terminal area should be used. Clearance should be provided between the termin[al] area and ground plane for two to fo[ur] connections to preserve circuit continuit[y]. See Figure 12-5; This prevents "heat sin[k]ing" of the terminal area during solderi[ng] operations; this can result in an inferi[or] solder joint.

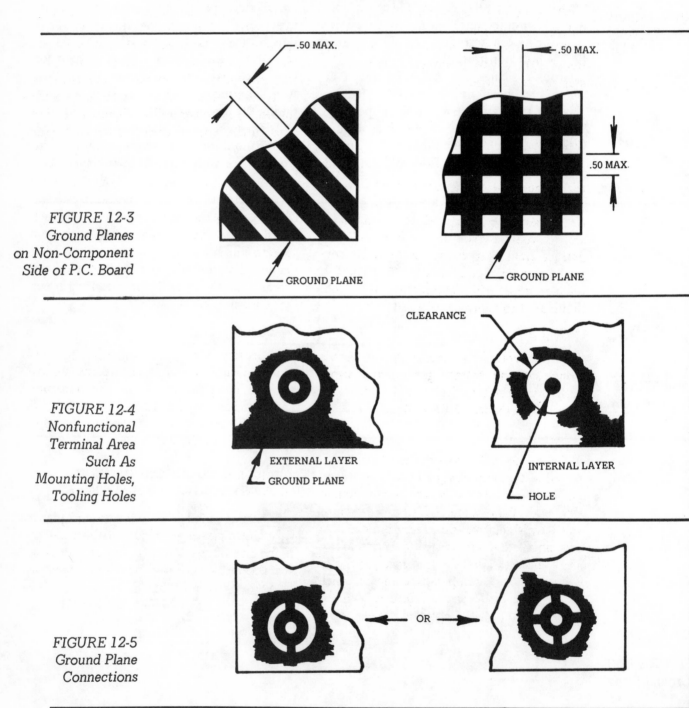

FIGURE 12-3
Ground Planes
on Non-Component
Side of P.C. Board

FIGURE 12-4
Nonfunctional
Terminal Area
Such As
Mounting Holes,
Tooling Holes

FIGURE 12-5
Ground Plane
Connections

GROUND PLANES

There are two methods used to construct voltage/ground plane. One is hand cutting and the second is photographically produced. These two techniques are itemized step-by-step below.

HAND CUTTING GROUND PLANES AND VOLTAGE PLANES

Hand cutting planes is accomplished with a material called Para-Paque* which is applied to the finished P.C. board, artmaster. Para-paque has an adhesive back to it and is applied over the surface and you extract areas that do not want to be covered by voltage or ground planes. The following steps will demonstrate the technique used to apply Para-Paque.

Step # 1

Apply your crop marks and all pads along with all voltage ground and signal paths used on a typical double sided P.C. board. (See Figure 12-6A & B)

Step # 2

Once the 2:1 artwork has been completed as far as applying all pads and traces, you now completely cover the component side with Para-Paque, by simply brushing and burnishing over the traces and pads. At this point you are ready to start cutting away the areas that do not tie to ground or voltage. Using a drop bow compass which has a knife blade inserted into it, you cut circles and extract the material away from

*Trade mark.

both the pads that do not connect and the pads that do strap to the plane. Once the clearance has been cut with the drop bow, simply take your x-acto knife and lift out the para-paque. Now take your tape, typically .100 tape, and strap the pad to the plane as demonstrated in Figure 12-5. For this cutting sequence see Figure 12-7 A, B.

Step # 3

To remove large sections from the center of the artwork or to cut long straight edges for clearance around the edge of the artwork use a straight edge along with your x-acto knife. The drop bow is only used to cut circles around your pads or to clear out areas for tooling holes or mounting holes. (See Figure 12-8 A, B)

Step # 4

This step demonstrates the final product which would be the finished plane applied to the component side or the non-component side. Figure 12-9 A, B shows the finished ground plane applied to the component side of the P.C. board.

MANUAL/HAND CUTTING SMALL ISOLATED VOLTAGE/GROUND PLANES

Creating small islands for ground planes or voltage planes or major bus lines can be done one of two ways. It can be cut using para-paque or it can be hand taped as shown in Figure 12-10.

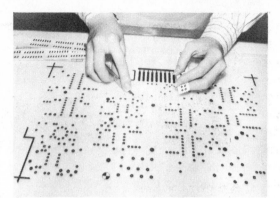

FIGURE 12-6A
Apply All Pads

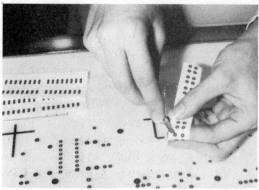

FIGURE 12-6B
Apply Crop Marks and All Pads

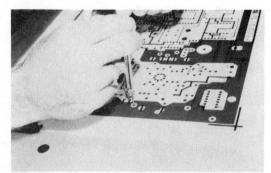

FIGURE 12-7A
Cut Clearance With Drop Bow

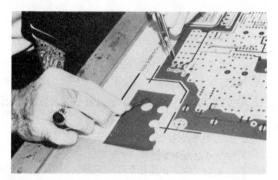

FIGURE 12-7B
Use X-Acto Knife to Remove
Para-Paque

FIGURE 12-8A
Drop Bow for Circles Only

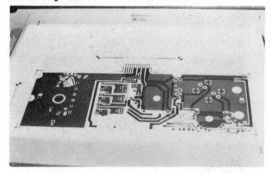

FIGURE 12-8B
X-Acto Knife for Straight Edge

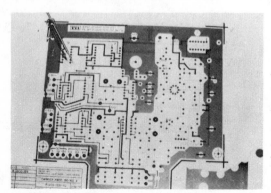

FIGURE 12-9A
Finished Ground Plane

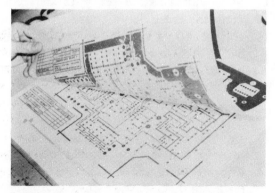

FIGURE 12-9B
Finished Ground Plane

Ground planes and voltage planes can be created using red para-paque along with blue and red tape. Figure 12-11 demonstrates the use of para-paque creating large massive areas for ground or voltage along with using red and blue tape. The main difference on red and blue as opposed to taping or cutting para-paque over black crepe tape is that using red and blue the tape for your traces is applied last rather than first.

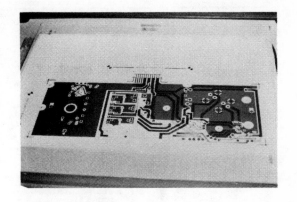

*FIGURE 12-11
Using Red Para-Paque
on Red and Blue
Artmasters*

*FIGURE 12-10
Creating Small
Islands for Ground
Plane*

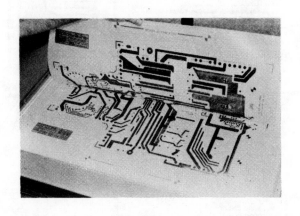

**PHOTOGRAPHICALLY PRODUCED
GROUND/VOLTAGE PLANES**

Ground/voltage planes can be produced photographically which eliminates the costly time consuming technique which was just demonstrated to you called manual/handcutting technique.

Care must be taken by the designer during his initial layout to assure one continuous ground plane. After the board is taped, a blue print of the component side of the artmaster can be used to show the photographer the ground plane limitations, and connections to be made. This print shall be sent to the photographers along with the artmaster and will be used to create the ground plane. See Figure 12-12.

NOTES FOR THE PHOTOGRAPHER–(THESE NOTES ARE TO BE PLACED ON OR ACCOMPANY THE BLUE PRINT.)

1. Ground plane side shown only
2. Air gap to be .XXX at 1/1 scale
3. Connect all indicated pads to ground plane
4. Remove all isolated grounds
5. Remove ground plane limitation line from finished neg. and positive

GROUND PLANES

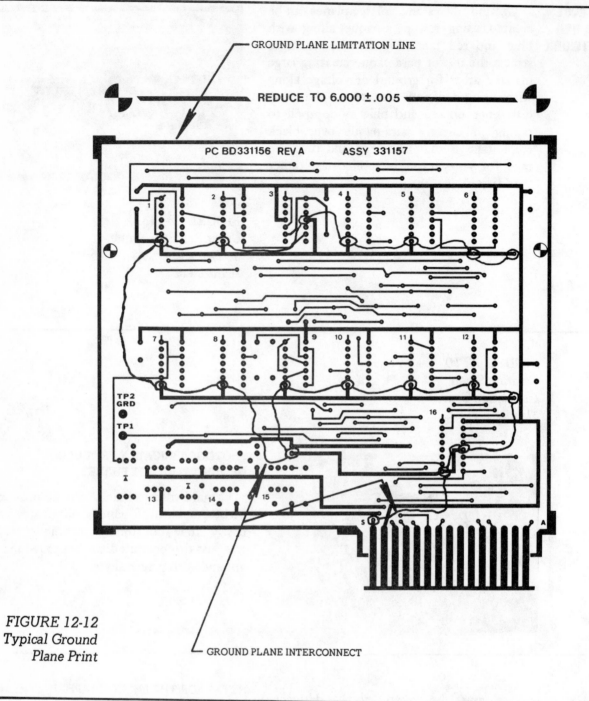

GROUND PLANE LIMITATION LINE

REDUCE TO 6.000 ± .005

PC BD 331156 REV A   ASSY 331157

TP2
GRD

TP1

FIGURE 12-12
Typical Ground
Plane Print

GROUND PLANE INTERCONNECT

Chapter

13

FAMILIARIZATION WITH FABRICATION DRAWINGS

The detail fabrication drawing of the printed circuit board shall depict the dimensional configuration of the board, size and location of holes, and shall specify the material, process specifications and (if applicable) notes and other information necessary for the fabrication of printed circuit board. Figure 13-1 shows a typical P.C. board detail drawing; detailed considerations are given below.

The detail fabrication drawing should be drawn in the same scale as the film or photomaster (normally). If more than one scale is used, each scale shall be clearly identified.

All dimensions needed to fabricate the P.C. board should be shown with the exception of those features located by the film or photomaster, such as pads, traces, connector fingers, etc. Dimensions for board configuration should be referenced from the tooling holes or register marks etched on the printed circuit board. All holes, cutouts, and the board outline shall be dimensioned from these points.

The detail fabrication drawing should show the non-component side of the P.C. board. One reason is because the manufacturer would like to view the side of the board that is drilled. Another reason is because of the burr left on the opposite side of drilling after the drill passes through the P.C. board.

The purpose of having the burr on the component side of the P.C. board is because widespread use of wave or flow soldering. This process requires that the side passing through the wave of solder be smooth, and the non-component side must be the side that passes through or over the wave of solder.

Manufacturing information, including plating and process specifications, shall be noted on the detail fabrication drawing (See examples below). Figure 13-2 shows a number of examples.

1. Hole sizes are after plating.
2. Circuit side of card shown.
3. Printed circuit board must conform to XXXX specification (See plating specifications below).
4. Printed circuit per photomaster XXXX, Revision X.
5. Hole diameter tolerance $^{+.005}_{-.002}$

The hole schedule chart appears on the face of the detail drawing and will give the symbol used to identify each hole called out. It contains a description of each hole size and tolerance, and if necessary, the quantity of each type or size used.

Listed below are most of the basic plating specifications requirements that must be provided to the P.C. board manufacturer. These specifications will not conform to all companies' standards; therefore, use these specs where applicable.

1. *Base Laminate:* Base material shall be laminated epoxy glass type G10, .06 ±.002 thick.

2. *Conductor:* Conductor material on double-sided (plated through) board shall be 1 oz. or 2 oz. copper minimum before plating.

3. *Solder Plating:* The tin lead alloy plating shall contain 40% ±5% lead and 60%

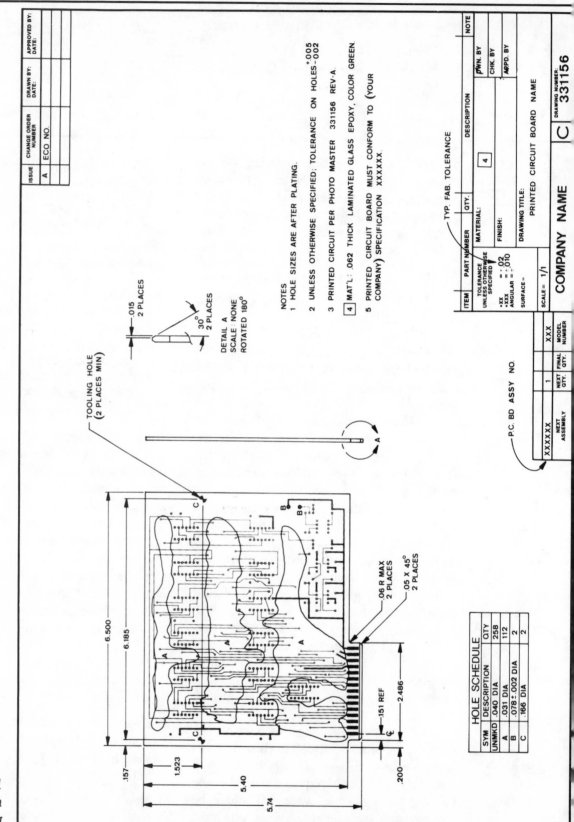

FIGURE 13-1
Typical Fabrication
Drawing

±5% tin. All solder-plated areas shall be subject to hot oil solder reflow process.

4. *Minimum Plating Thickness on Trace Surface and Through Holes* (inches):

Copper   .001
Tin-lead   .0003
Gold   .00005
Nickel-gold   .0003

5. *Minimum Plating Thickness for Fingers* (inches):

Nickel-gold   .0003
Nickel-rhodium   .0003
Gold   .00005

6. *Conductor Defects:*

   a. Pin holes, nicks, and scratches: Conductor defects such as pin holes and nicks which do not reduce the conductor width by more than ±.002-in.

   b. Plating overhang: The plated resist material overhang remaining after the copper foil has been removed by etching shall not exceed .005-in.

   c. Undercut: The reduction of conductor caused by etching under the edge of plating resist shall not exceed .005-in.

   d. Excess conductor material: Isolated spots of conductor material no less than .015-in. in diameter are acceptable providing spacing requirements are not violated.

7. *Warpage:* Warpage or twist of a printed circuit board shall not exceed .005-in. per inch.

8. *Solderability:* The solder coating shall be adherent and non-granular in appearance. The base material shall not show evidence of blistering or delamination, nor shall conductors or terminal areas show evidence of separating from the base laminate after dip soldering.

1. Hole sizes are after plating.
2. Circuit side of card shown.
3. Printed circuit board must conform to **XXX** specification.
4. Printed circuit per photo master, **XXX**, Rev **XXX**.
5. Hole diameter tolerance = $+.005 \atop -.002$

Hole Schedule		
Symbol	Description	Qty.
Unmarked	.038 Dia.	105
A	.048 Dia.	59
B	.187 Dia.	7

TABLE 13-1

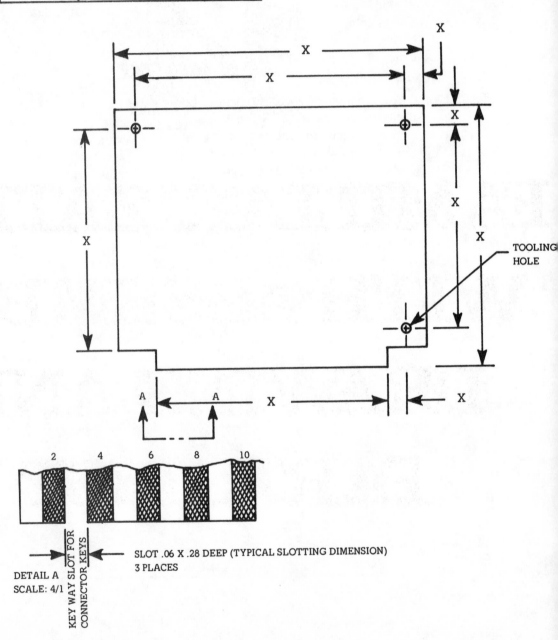

TOOLING HOLE

SLOT .06 X .28 DEEP (TYPICAL SLOTTING DIMENSION)
3 PLACES

DETAIL A
SCALE: 4/1

KEY WAY SLOT FOR CONNECTOR KEYS

FIGURE 13-2
Notes and Hole
Schedule

Chapter

14

FAMILIARIZATION WITH ASSEMBLY DRAWING AND PARTS LIST

The printed circuit assembly drawing should meet the normal requirements of an assembly drawing. It should depict the P.C. board components, reference designation markings, assembly and test specs, and also include a list of material which contains all part information needed to order the parts. A typical assembly drawing is shown in Figure 14-1; detailed considerations are given below.

SCALE

The assembly drawing is normally drawn in the same scale as the artwork or tape-up. If more than one scale is used, each scale shall be clearly identified.

VIEW

The assembly drawing shall be drawn as viewed from the component side. Also a side view shall be shown to indicate the maximum component height relative to the board. Enlarged views or sections may be used if needed for clarification.

SYMBOLS & REFERENCE DESIGNATIONS

Symbols and reference designations should be shown per the applicable schematic diagram. Abbreviations, if used, shall be per ANSI USASI-Z32.13 or applicable company standards. Reference designations should also be used to identify all electrical components from the assembly drawing back to the schematic and parts list, (Figure 14-2).

MOUNTING

Methods for mounting the components, such as riveting, soldering, brackets, etc., shall be noted.

JUMPERS

If wire terminations are used for circuit interconnections and are not etched or marked on the P.C. board, the wire terminations shall be identified as jumper wires and shown on the assembly drawing.

ORIENTATION & INDEXING

All transistor tabs should be shown unless it is an epoxy-type transistor, which will be indicated by the flat side. TO-18 and TO-5 case pins 8 or 10 should be marked. For dual-in-line cases, pin 1 should be marked. All polarized capacitors (plus side) and all diodes (cathode end) should also be marked.

NOTES

Sample P.C. assembly drawing notes, as applicable, are listed below.

1. Reference electrical schematic diagram XXXXXXX.

2. Assemble and solder per XXXX spec (or XXXXX standard practice).

3. Observe polarity of capacitors, diodes, etc.

4. Mark assembly number and revision letter (per applicable company standard).

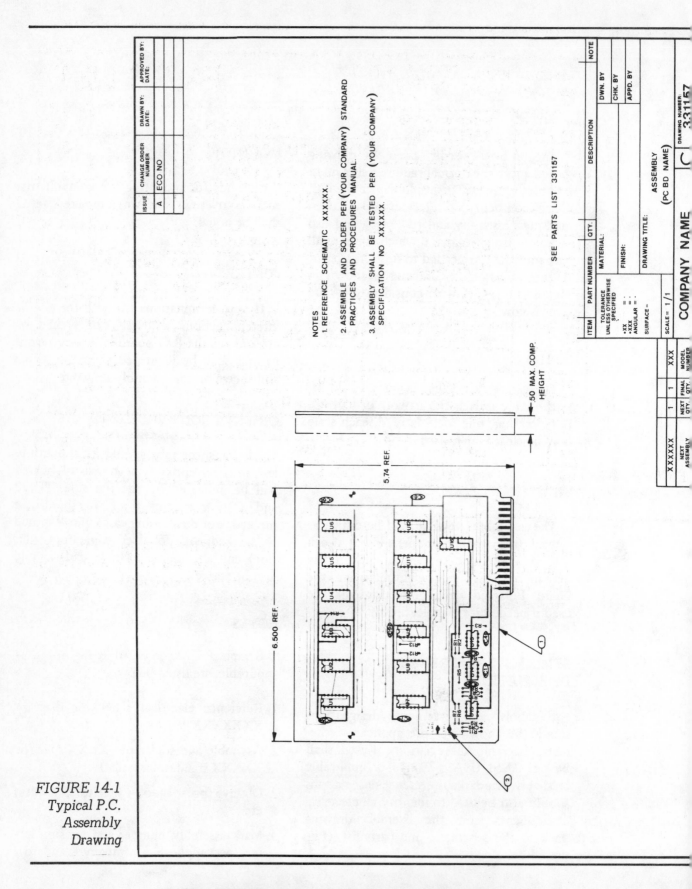

FIGURE 14-1
Typical P.C.
Assembly
Drawing

FAMILIARIZATION WITH ASSEMBLY DRAWINGS

REF DES	ITEM NO.	MANUFACTURERS PART NUMBER	DESCRIPTION	TOTAL QTY
C1		CKO5	CAPACITOR, CERAMIC 0.1 UF, 10V	3
C2		CKO5	CAPACITOR, CERAMIC 0.1 UF, 10V	-
C3		CKO5	CAPACITOR, CERAMIC 0.1 UF, 10V	-
R1		89PR5K	RESISTOR, VARIABLE 5K	1
R2		JANRCRO7	RESISTOR, CARBON 560, 1/4W, 5%	2
R3		JANRCRO7	RESISTOR, CARBON 4 7K, 1/4W, 5%	1
R4		89PR10K	RESISTOR, VARIABLE 10K	1
R5		JANRCRO7	RESISTOR, CARBON 560, 1/4W, 5%	-
Q1		2N3210	TRANSISTOR NPN	2
Q2		2N3210	TRANSISTOR NPN	-
U1		LM741	OP-AMP	2
U2		LM741	OP-AMP	-

ASSEMBLY NAME POWER SUPPLY
ASSEMBLY NUMBER 331578

PARTS LIST

FIGURE 14-2
Parts List

5. Board to be conformal coated in accordance with XXXX MIL Spec. Test points and adjustable components shall be masked during coating.

6. Electrical reference designations are for reference only and need not appear on the parts or components unless otherwise specified.

Chapter

15

SILKSCREEN DRAWINGS AND ARTMASTERS

The silkscreen drawing and artmaster provide a tool from which silkscreens, stamps, stencils or other marking mediums are fabricated for marking the printed circuit board with reference markings prior to assembly of the components.

SILKSCREEN ARTMASTER

The artmaster seen in Figure 15-1 should have sharp definition and high contrast because it is reproduced by a photographic process.

The silkscreen artmaster should be a transparent, stable material either with black self-sticking printed circuit tape, black matte characters having a heat resistant, pressure sensitive adhesive backing or black ink lettering with a "Leroy" or similar type pen. The tape and pre-printed characters are preferred because of their well defined lines and uniform contrast.

Ragged edges, splits on the line work or characters, and pinholes should be avoided in order to reduce the amount of retouching.

The artmaster should be made to the same scale as the printed circuit artmaster.

The silkscreen artmaster is an undimensioned master. The artmaster should show the accurately scaled distance between two target points in the horizontal or vertical planes; both horizontal and vertical are not necessary, but may be used.

Tooling holes or fabrication lines for the artmaster should be located on the artmaster to correspond with those on the printed circuit artmaster.

All markings should be taken from the schematic or bill of materials (reference designations) of the printed circuit board. They should be located to cause no identification confusion.

SILKSCREEN DRAWING

The silkscreen drawing is an undimensioned drawing used primarily to denote color of silkscreen and special instructions, Figure 15-2.

The silkscreen drawing should be made to the same scale as the artmaster or reduced to fit the drawing size if the artmaster scale is too large. It should be drawn as viewed from the component side of the printed circuit board.

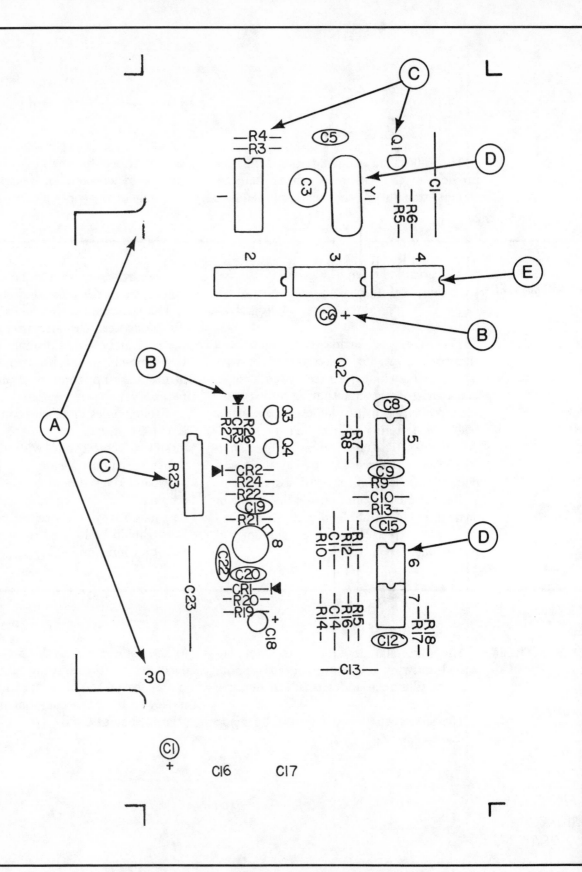

SILKSCREEN DRAWING AND ARTMASTERS

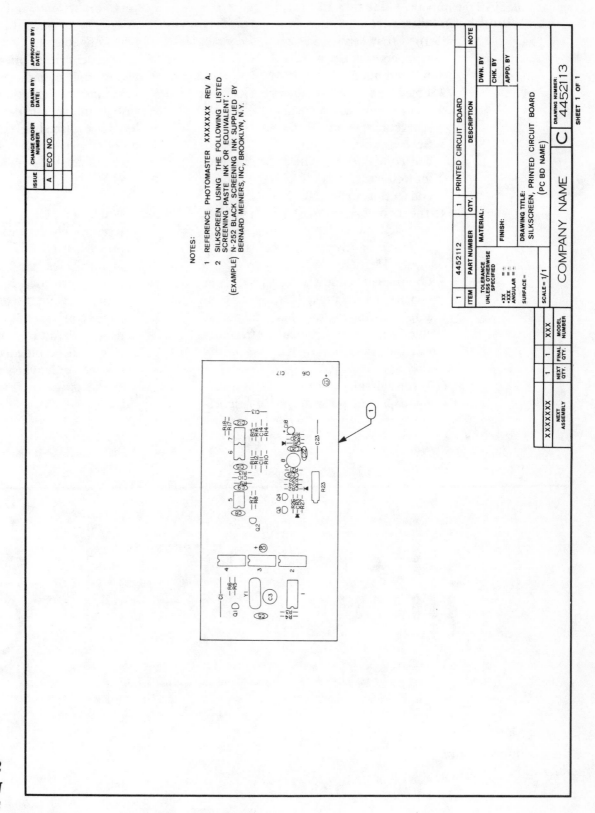

FIGURE 15-2
SILKSCREEN
DRAWING

The following items must appear on the component side of a PB: (capital letters in parentheses refer to item in Figure 15-1).

1. The first and last printed-contact numbers/letters, if board edge contacts are used. (A)
2. Etched component legends: Shall not reduce conductor spacing below design requirements. (B, C)
3. Screen-printed component legends and component outlines: Shall not be located on any lands or other solder connection areas. (D)
4. The component-orientation pin, tab, or mark on all components shall be indicated on the artmaster as shown in Figure 15-1. (E)
5. Component numbers for active components (transistors, IC's, operational amplifiers, power supplies) and adjustable passive components (trimpots, variable capacitors) shall be visible after assembly. (C)
6. Test-point identification, diode polarity, capacitor polarity, board title. (B)
7. Other component numbers (for resistors, capacitors, and small-signal diodes) should be visible after assembly when possible.
8. Component lead orientation identification shall be indicated when correct component orientation is not evident. For example, when in-line transistor leads are placed in radial component holes or vice-versa, an E, B, or C designation shall be included. (See Figure 15-3.)

1.5.3 When preparing the component legend, whether etched or screen printed, it is not necessary to block in small resistors, capacitors, and diodes, provided mounting-hole locations are unobstructed. However, large components, unique components, hardware, transistors, and IC's shall be blocked in for ease of loading (only when a screen-printed legend is used). (Figure 15-1, Item D).

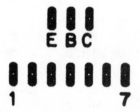

FIGURE 15-3
Component-orientation information

INSPECTION OF GENERAL PRINTED CIRCUIT BOARD MASTER DRAWINGS
P.C. Documentation inspection should include the items in the following checklist:

1.0   GENERAL REQUIREMENTS
    1.1   Do the title blocks of all drawings, all sheets include:
        1.1.1   Correct number and subnumber; drawing type and size?
        1.1.2   Complete, correct title?
        1.1.3   Correct number of sheets of artwork and proper callout ("Sheet 2 of 4" etc.)?
    1.2   Are abbreviations, component designations, spelling correct on all drawings and all sheets?

2.0   MATERIAL AND DRAWING LIST (PARTS LIST)
    2.1   Is the material description on the drawing complete?
        2.1.1   Is part identification complete?
        2.1.2   Is "quantity required" indicated?
        2.1.3   Is manufacturer's description complete?
        2.1.4   Are stock numbers listed?
    2.2   Are all reference drawings listed?
    2.3   Are all specifications listed?

3.0   PHYSICAL REQUIREMENTS
    3.1   Are all dimensions correct?
    3.2   Are tolerances included?
    3.3   Are required key slots included? Size and location?
    3.4   Are tooling holes included? (Standard on all artwork.)
    3.5   Are special mounting considerations (of heat sinks, etc.) noted?
    3.6   Are special plating requirements included?

4.0   COMPONENTS
    4.1   Correct value, physical size and lead size?
    4.2   Are component designation numbers correct? Are there any missing or duplicated items?
    4.3   Are polarities correct? (When applicable.)
    4.4   Are there any missing components?
    4.5   Is component location correct, per originator?

5.0   SILK-SCREEN
    5.1   Is the layout clean and neat?
    5.2   Are the number and subnumber included?
    5.3   Is the title included (if space is available)?

6.0   ARTWORK
    6.1   Are the number and subnumber included on both sides of artwork?
    6.2   Are there wrong connections?
    6.3   Conductors. Are there:
        6.3.1   Missing conductors?
        6.3.2   Missing jumpers?
        6.3.3   Gaps or notches in conductors?
        6.3.4   Lifted conductors?
        6.3.5   Is the minimum distance between conductors acceptable?
        6.3.6   Is the minimum distance between pads acceptable?

7.0   FABRICATION AND ASSEMBLY
    7.1   Does tabulated hole "quantity" match board?
    7.2   Are drill sizes correct for component lead diameters?
    7.3   Are pad sizes correct for special hardware or components?
    7.4   Is special component mounting information or manufacturer's informa-
          tion included in "Remarks"?
    7.5   Are special plating requirements included?
    7.6   Are special assembly requirements included?

Chapter

16

MANUFACTURING PROCESS OF P.C. BOARDS

ADDITIVE AND SUBTRACTIVE PROCESSES

This chapter introduces the techniques of pc board manufacturing, so that designers may appreciate its latitudes and its limitations and hopefully avoid the disastrous consequences of designing a board that cannot be manufactured.

Both subtractive and additive techniques are presented, and a series of photos detailing the entire manufacturing process is also given.

Conventional subtractive processes are currently the most popular manufacturing techniques. This approach begins with a pc board which is entirely coated with copper. PC board artwork described in previous chapters is used to construct an acid-resistant shadow mask on the copper—when submerged in acid, the copper is eaten away, leaving a pattern of conductive "wires" in the copper. This conductive pattern is the board's printed circuit.

But limitations with the subtractive process hint that it may be augmented (or even replaced) by an additive process which this chapter describes. Although more costly to use, the additive processes eliminates the time-critical risks of acid etching, and allows manufacturers to improve pc board circuit density and product yield.

The chapter includes a complete series of photos showing all the steps in a contemporary, computer-controlled manufacturing process.

Today, the printed circuit board market is touching every facet of life from computer electronics, to automobile, dishwasher, etc. And most of the boards being manufactured today are using a technique called subtractive process. This essentially is a 30-year old technology summarized in Table 16-1. There is now a new technology called "additive process," summarized in Table 16-2. Today there is a constant pressure throughout the entire electronics industry to provide more compact and denser P.C. boards. This pressure has brought about the use of solder masks which will be discussed further in this chapter.

The basic sequence for manufacturing P.C. boards by the conventional or subtractive process is shown on Table 16-1.

A. Copper clad base material.
B. Drill or pierce holes.
C. Sensitize holes.
D. Electroless copper flash.
E. Print reverse pattern mask.
F. Electroplate copper.
G. Electroplate etch resist metal.
H. Strip Mask.
I. Etch foil copper.
J. Apply solder mask (optional).

TABLE 16-1
Conventional
Subtractive Process

Starting with panels of copper clad base material, holes are fabricated by drilling or piercing. The panels are sensitized and given an electroless copper flash to make the holes conductive for electroplating. A plating resist mask is applied by screening or photoprinting and copper is pattern plated with the minimum thickness in the holes, usually at least .001-in. A second metal is plated over the copper to act as an etch resist and to provide corrosion protection for the finished circuit. The screened or photoprinted mask is stripped, the background copper is etched and a solder mask may be applied if specified by the end-user.

The basic sequence for manufacturing P.C. boards by the additive process is shown on Table 16-2.

TABLE 16-2 *New Additive* *Process for Greater* *Control*	A. Adhesive coated unclad laminate. B. Drill or pierce holes. C. Print reverse pattern mask. D. Activate. E. Electroless copper deposition. F. Strip Mask (optional). G. Apply solder mask (optional). H. Apply protective overcoat.

"A STEP-BY-STEP PROCESS THROUGH A TYPICAL P.C. MANUFACTURING FACILITY"

FIGURE 16-1

STEP 1: Film is inspected by the P.C. board manufacturer.

FIGURE 16-2

STEP 2: The board manufacturer select: panels of copper-clad base material.

FIGURE 16-3

STEP 3: Holes are drilled in the board. In order to use a numerical control drilling machine, a program tape must be generated. Figure 16-3 shows a tape-generating machine.

FIGURE 16-5

FIGURE 16-4

STEP 4: The tape machine registers the pad coordination in the X and Y direction.

STEP 5: The pads are located optically for the XY directions (Figure 16-5).

Using this method, all holes are located with reference to pad centers. This method permits the use of drilling holes located directly from the artmaster or film.

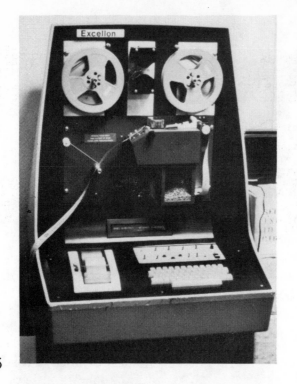

FIGURE 16-6

STEP 6: Figure 16-6 shows finished tape ready to be used for the drilling sequence.

FIGURE 16-7

STEP 7: The tape is loaded into the computer which will control the drilling machine.

FIGURE 16-8

STEP 8: Shows typical drilling machine. The machinery used for drilling the holes i positioned by numerically controlled tape The air-driven motors vary from 75,000 tc 125,000 RPM with a driving pressure of 500 PSI. After drilling, the boards are deburred by sanding.

FIGURE 16-9

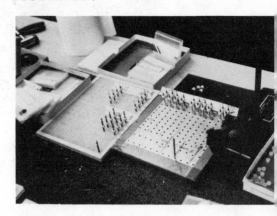

STEP 9: Shows the selection of drill which will be changed for each tape sequence.

FIGURE 16-10

FIGURE 16-12

STEP 10: The panels are sensitized and given an electroless copper flash to make the holes conductive for electroplating.

STEP 12: A plating-resist mask is applied by screening or photo-printing, The silkscreen method uses 250 or 165 mesh stainless steel screens. In the silkscreening process, the screen is painted with a red emulsion. The emulsion is then photographically exposed through the negative with ultra-violet rays. These rays make the exposed areas of the emulsion harden. The screen is processed, leaving only the hardened areas—thus producing the image.

FIGURE 16-11

FIGURE 16-12A

STEP 11: Copper is plated with the minimum thickness in the holes, usually at least .001 in.

Figure 16-12A shows a screening room.

FIGURE 16-12B

Figure 16-12B shows the panel being positioned under the silkscreen.

FIGURE 16-12C

The resist (blue in color) is coated on th
screen and squeegeed through the scree
onto the panel.

Figure 16-12D shows the result of the screening process.

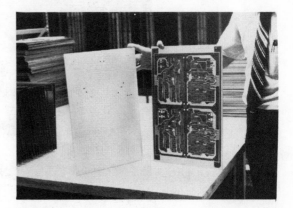

FIGURE 16-14

STEP 14: After the panels are screened and baked an inspection process is performed before plating.

FIGURE 16-14A

FIGURE 16-13

STEP 13: The panels are baked in the ovens.

MANUFACTURING PROCESS OF P.C. BOARDS   267

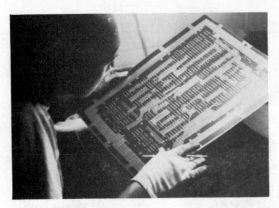

FIGURE 16-16

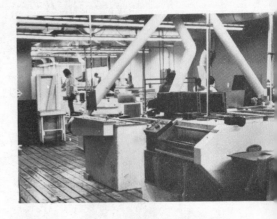

FIGURE 16-14B

STEP 16: The screened or photoprinted mask is stripped. The background copper is etched off. Figures 16-16 and 16-16A show a typical etching machine.

FIGURE 16-16A

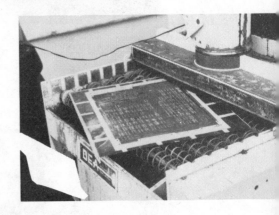

FIGURE 16-15

STEP 15: A second metal is plated over the copper to act as an etch resist and to provide corrosion protection for the printed circuit board.

Some of the more popular metals are: tin, lead, gold, nickel-gold, nickel-rhodium.

FIGURE 16-17

STEP 17: Cutting to Final Size and Shape. To cut the board to the final size requires a metal template that has been prefabricated from the fabrication drawing to the precise outer dimensions of the P.C. board.

FIGURE 16-18

STEP 18: A routing tool is used along with the template to cut the board outline, slots, holes, or cutouts. Figures 16-18, 16-18A, and 16-18B show the routing machine.

FIGURE 16-18A

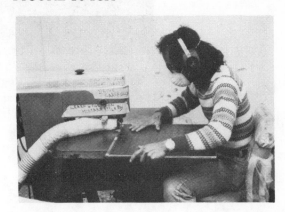

MANUFACTURING PROCESS OF P.C. BOARDS   269

FIGURE 16-19A

FIGURE 16-18B

FIGURE 16-19B

FIGURE 16-19

STEP 20: Wrap, bag and ship back customer.

STEP 19: Final Inspection. Figures 16-19 and 16-19A: Visual inspection under magnification. Figure 16-19B: Electronically checking for plating thickness.

Solder masking is a coat of epoxy resin, covering all areas of the P.C. board except pads which require soldering. See Figure 16-20.

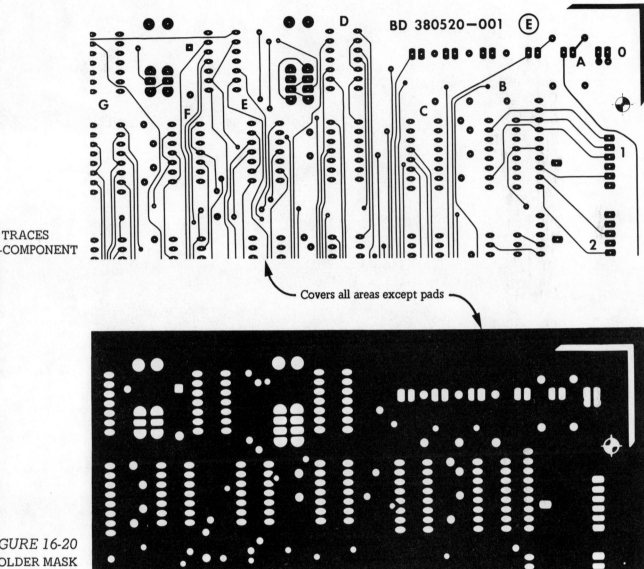

TRACES
COMPONENT

Covers all areas except pads

FIGURE 16-20
SOLDER MASK

With the current industry trend towards higher density boards with tighter line widths and spacing, solder masks are increasingly being used to eliminate bridging between adjacent conductors during wave soldering.

Wave soldering is a process whereby component leads are soldered to a printed circuit board by traveling at a predetermined rate of speed through a wave of molten solder.

Chapter

17

FLEXIBLE CIRCUITS

This chapter introduces flexible printed circuit boards for use in electronic systems.

Sometimes, flex circuits are used as 3-dimensional pc boards, bent to fit into a specifically-shaped volume, or to serve as connecting cables between subsystems. Other times, they are used to provide power and logic to electronic parts that move on hinges, like computer face panels, or ones that move on rails, like the dot-matrix print heads on contemporary hard-copy printers.

Flex circuits have been called "new" for a quarter-century or more, and recent advances in electroless copper, insulating films and adhesives promise to continue this tradition. As costs erode and quality and reliability increase and new applications arise, more and more pc board designers will be faced with the task of designing stretchable, bendable and foldable printed circuit boards.

This chapter shows how to deal with the looser tolerances and dimensional instability of flex circuits, and it describes the four basic requirements that designers must deal with, the five ways to lay out and manufacture flex circuits and the important techniques that shield flexible conductors from interfering with one another.

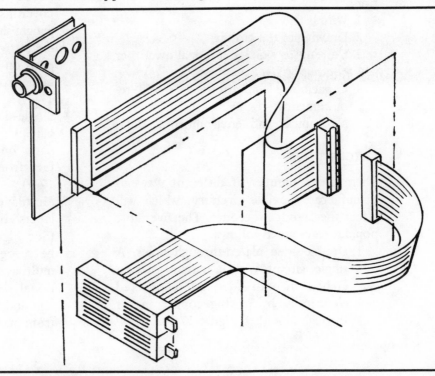

FIGURE 17-1

DESIGN AND FABRICATION

In specification as well as manufacturing, flexible printed circuits bear little resemblance to rigid printed circuitry. Warp and flatness specifications are meaningless because of the flexible nature of the product. Tolerance regarding finished circuit dimensions are looser due to the dimensional instability of the insulation films as compared to rigid boards. By the same toke flexible circuitry can be stretched, bent, folded, compensating somewhat for tl looser tolerances in flex circuits. These ar other design and manufacturing consider tions demand that circuit designers unde stand the basic requirements of flexible c cuits as opposed to rigid circuit design.

PRINCIPAL DESIGN CONSIDERATIONS

THERE ARE FOUR PRINCIPAL CONSIDERATIONS BEHIND THE DESIGN OF FLEXIBLE CIRCUITS.

1. COST: The total installed cost. Not simply the price of the flex circuit compared with wire.

2. FUNCTION: An evaluation of the circuit's purpose and whether that purpose could be served as well by going to flex circuit.

3. WIRING: Mass circuit terminatio could production time be saved a wiring errors be eliminated.

4. COMPLEXITY: If total quantities a relatively small, the circuit must very complex to be economica worthwhile. In large quantities, ho ever, simple as well as complex conf urations will pay for themselves.

FLEXIBLE CIRCUIT LAYOUT

LAYOUT REQUIRED INFORMATION

Almost all flexible circuits are custom designed. Designers must have this information:

1. Wire list.
2. Drawing of the circuit.
3. Current or voltage drop allowed per conductor.
4. Capacitance limitations or shield requirements.
5. Mechanical and environmental data.

LAYOUT TYPES

There are a number of different ways to manufacture flexible circuitry, which will affect the layout procedure. The five most popular ways are as follows:

1. Double-sided plated-through holes. A single circuit with a circuit pattern etched on both sides, interconnected electrically by plating copper in the through hole (See Figure 17-2).

2. Single-sided. A single-sided circu with a single pattern etched on o side only, but with solder pads e posed through both the base an cover insulation. (See Figure 17-3).

3. Single-sided—a normal, single-side circuit which simply places the sold pads on the one side and allows co tact from either side (see Figu 17-4).

4. Single-sided—a normal single-sided ci cuit which simply places the sold pads on one side and allows co tact from only one side (see Figu 17-5).

5. Multilayer—three or more conductiv layers limited only by economic (See Figure 17-6 and 17-7). It shoul be recognized that for each layer c multilayer structure there is add tional dedicated manufacturing too ing and a cumulative cost resultin from manufacturing yields.

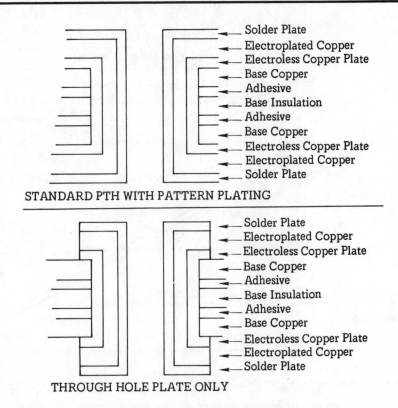

Solder Plate
Electroplated Copper
Electroless Copper Plate
Base Copper
Adhesive
Base Insulation
Adhesive
Base Copper
Electroless Copper Plate
Electroplated Copper
Solder Plate

STANDARD PTH WITH PATTERN PLATING

Solder Plate
Electroplated Copper
Electroless Copper Plate
Base Copper
Adhesive
Base Insulation
Adhesive
Base Copper
Electroless Copper Plate
Electroplated Copper
Solder Plate

FIGURE 17-2

THROUGH HOLE PLATE ONLY

This type of flexible printed wiring is made up of three layers—two dielectric layers and a single conductor layer, see Figure 17-3. The conductor layer is sandwiched between the two insulating layers which have selective access holes to the conductor. The access holes may be at opposing points leaving the bare conductor exposed on both sides, or the access areas may be selectively located on either side.

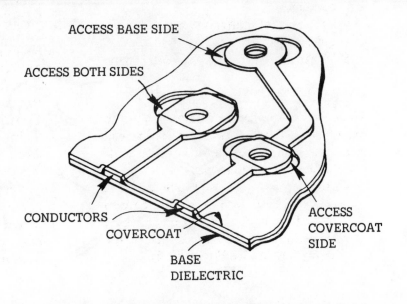

ACCESS BASE SIDE
ACCESS BOTH SIDES
CONDUCTORS
COVERCOAT
BASE DIELECTRIC
ACCESS COVERCOAT SIDE

FIGURE 17-3

This type of flexible printed wiring has the unique feature of electrical access to the single layer conductor from both sides. The structure consists of two layers, dielectric and conductor; Figure 17-4. The conductor pattern is suspended across the holes allowing contact from either side of flexible printed wiring.

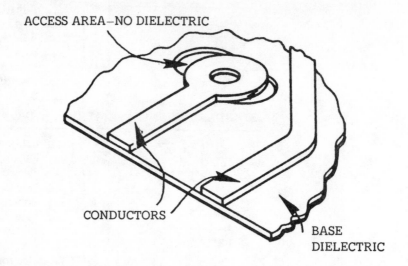

ACCESS AREA–NO DIELECTRIC

CONDUCTORS

BASE
DIELECTRIC

FIGURE 17-4

This form of flexible printed wiring finds most application where the wiring will not be exposed to mechanical abuse or environmental contamination. The components are normally all mounted from the dielectric side with all electrical connections to the conductor side. The connections can be soldered, welded or swaged type. This type of wiring has found great acceptance in such items as telephones, relays, toys, small motors and many other such assemblies which are small enclosed functional items. See Figure 17-5.

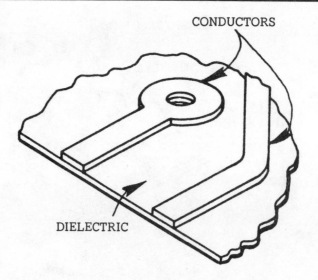

CONDUCTORS

DIELECTRIC

FIGURE 17-5

The multilayer structure in its simplest form (Figure 17-6) is a three conductor layer with the center layer being discrete wiring paths and both outer conductor layers a solid copper conductor. In this type of multilayer the wiring paths are electrically equivalent to co-axial or shielded wire configurations.

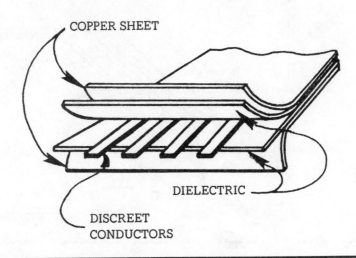

*FIGURE 17-6
Simple Multilayer
Flexible Printed
Wiring Structure*

COPPER SHEET

DIELECTRIC

DISCREET
CONDUCTORS

The most frequently used structure of multilayer construction is a four layer structure (Figure 17-7) utilizing the adjacent inner layers for power and ground distribution and the outer layers for discreet point to point interconnection, plated through holes provide the interconnection between layers.

Many designers have found that in utilizing this concept, a significant savings in cost and manufacturing can be achieved. This is accomplished by a standard configuration of the ground plane and power plane layers. The fabrication of the signal plane paths and interconnection between layers is then a process equivalent to fabricating standard through-hole plated two sided printed wiring board.

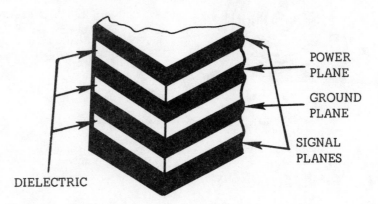

*FIGURE 17-7
Four Layer Flexible
Printed Wiring
Structure*

POWER
PLANE

GROUND
PLANE

SIGNAL
PLANES

DIELECTRIC

Right Angle Fold Figure 17-8.
This type of fold technique is specifi-

cally for a non-moving application whe
there is no demand for flex-life.

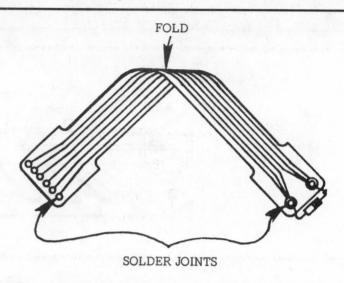

FOLD

SOLDER JOINTS

FIGURE 17-8

MOVING/ROLLING Figure 17-9.

The prime application for a moving or rolling type of flex printed circuit (Figure 17-9) is a roll-up, which could attach from a sliding drawer. The type of dielectric

material is limited to attain the windo
shade effect—to those dielectrics which ca
be treated to create the memory for rol
up.

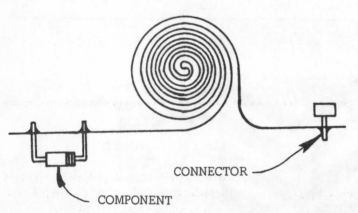

CONNECTOR

COMPONENT

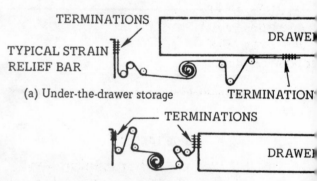

TERMINATIONS

TYPICAL STRAIN
RELIEF BAR

DRAWE

(a) Under-the-drawer storage

TERMINATION

TERMINATIONS

DRAWE

(b) Behind-the-drawer storage

FIGURE 17-9

FLEXIBLE CIRCUITS

ACCORDION

The Accordion Technique (shown in Figure 17-10 and 11) can be used with dielectric and materials which have a weaker spring action. This type of application usually requires some form of external support member to control its position in the retracted condition.

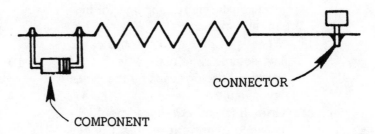

CONNECTOR

COMPONENT

FIGURE 17-10

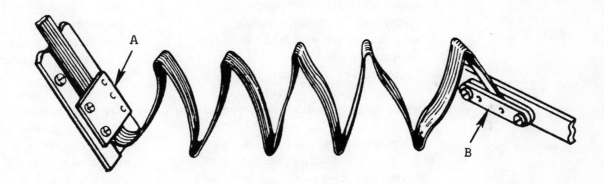

A

B

FIGURE 17-11
Accordion cable
with stress relief
at points A and B.

LAYOUT STEPS

1. Work at an enlarged scale.
2. Place the various connectors or components to be wired in their exact position.
3. Determine all physical restrictions.
4. Using the schematic, interconnect the various points.
5. Avoid crossovers if possible. Arrange the wiring points so that a simple one-for-one wiring plan is achieved.
6. At this time, it will be evident if there is a need for cross-overs which may require a multilayer circuit.

SPECIAL DESIGN CONSIDERATIONS

1. For structural strength and heat sinking, copper should be maximized wherever possible.
2. Conductor runs parallel to fold lines could make folding difficult and should be avoided. (Figure 17-12.)
3. Pads should be filleted for added strength during folding and for heat sinking properties (Figure 17-13.)
4. Where possible, artwork and coverlay designs should provide for the capture of solder pads by the coverlay. (Figure 17-14.)
5. Crop lines or cut lines should be radiused to prevent tearing. Copper-reinforced tear-stops are also recommended.
6. Spiral conductors are more easily defined and require less time to tape if defined as concentric circles connected by tangents rather than as a circular curve with a constantly increasing radius.
7. Part numbers, if required, should be on the tape-up to allow them to be etched into largest available copper area.
8. When taping narrow traces, remember that a 20% allowable reduction from artwork-to-finished part in conductor widths and spaces is often required. It is very desirab to have conductor widths and a gaps of .050 in. or greater.
9. Visualize the positioning of the ci cuits on a roll of material. As th circuits are positioned adjacent t each other on the roll, any waste material between the circuits adds cost which must be incorporated i the final price of the finished ci cuit board.
10. Pads should be at least .030 in larger in diameter than the throug hole diameter to allow for misregi tration tolerance, wherever desig permits.
11. There are many methods whic may be employed to create an in terconnection between flexible ci cuitry and other adjacent compo nents. Some of these methods ar solderless and some use solder. Al though the solderless technique may appear to be more expensiv because of the initial cost of th connecting device, they are usuall easy to assemble, and lend them selves well to assembly line or auto mated processing. However, flexibl circuit may also be wave soldere and hand soldered.

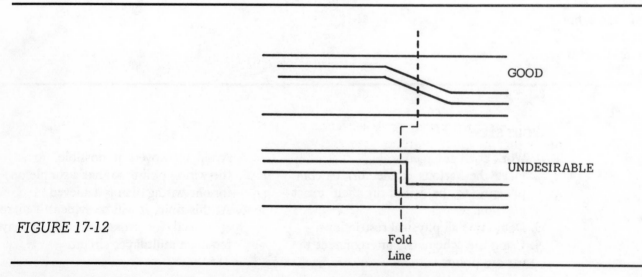

GOOD

UNDESIRABLE

FIGURE 17-12

Fold
Line

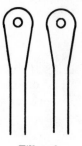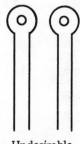

Filleted
Pads

Undesirable
Non-filleted
Pads

FIGURE 17-13

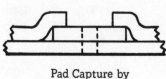

Pad Capture by
the Coverlay
Insulation

FIGURE 17-14

CONDUCTOR SPACING

Most often the spacing is chosen because of manufacturing considerations rather than by electrical performance.

Voltages are normally low and most materials used can withstand 300 V or more. Standard flex-circuits feature air gaps from .010 to .050 in. When required, they can be as small as .003 in.

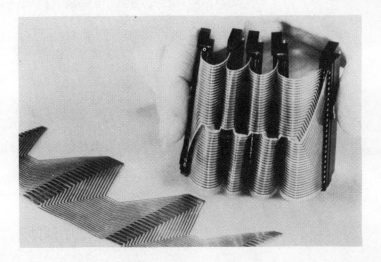

CONDUCTOR SIZE

Most interconnections carry currents in the milliamps range, but there are requirements to carry greater currents. Usually, these are expressed in "AWG" for a conductor cross-section.

Example—31 AWG Wire: The flat conductor equivalent for 1 oz. copper would be .0473 in. in width (use .050 in.) 2 oz. copper = .025 in. in width.

If 30°C is used to determine current-carrying capacity, it may be possible to up-rate the AWG requirement because flat conductors in flex-circuits will dissipate heat more efficiently. See Table Figure 17-15.

FLEXIBLE CIRCUITS

COPPER CONDUCTOR CHARACTERISTICS

Flat Conductor Dimensions		Cross Section		Nearest Equiv. (based on equiv. cross section)	A.W.G. Wire Size (based on Equiv. current rating)	Resistance (milliohms per ft. at 20°C	Current (amp for 30°C rise
Thickness (in)	Width (in)	(sq. mils)	(circular mils)				
.0027	.030	81	102	30	28	100	3.4
.0027	.045	122	154	28	27	67	3.8
.0027	.060	162	204	27	25	50	5.1
.0027	.075	202	254	26	24	40	5.8
.0027	.090	243	306	25	23	34	6.5
.0027	.125	338	425	24	22	24	8.2
.0027	.155	418	527	23	21	19.5	9.2
.0027	.185	500	630	22	20	16.2	10.7
.0027	.250	675	850	21	18	12	13.5
.004	.030	120	151	28	27	67	4.0
.004	.045	180	227	26	25	45	5.2
.004	.060	240	302	25	24	34	6.0
.004	.075	300	378	24	23	27	7.0
.004	.090	360	454	23	22	22.5	7.8
.004	.125	500	630	22	20	16.2	10.0
.004	.155	620	780	21	19	13	11.8
.004	.185	740	930	20	18	11	13.5
.004	.250	1000	1260	19	17	8	17.0
.0055	.045	248	312	25	24	33	6.0
.0055	.060	330	415	24	23	25	7.2
.0055	.075	412	520	23	22	20	8.2
.0055	.090	495	624	22	21	16.5	9.5
.0055	.125	687	865	21	20	12	12.2
.0055	.155	852	1075	20	19	9.5	14.8
.0055	.185	1020	1285	19	17	8	17
.0055	.250	1375	1730	18	16	6	21
.008	.045	360	454	23	22	23	7.8
.008	.060	480	605	22	21	17	9.8
.008	.075	600	755	21	20	13.5	11.5
.008	.090	720	905	20	19	11.2	13.2
.008	.125	1000	1260	19	17	8	17
.008	.155	1240	1560	18	16	6.5	20
.008	.185	1480	1860	17	14	5.5	23
.008	.250	2000	2520	16	13	4.1	26

FIGURE 17-15

FLEXIBLE CIRCUITS

PAD AREA

Make pad area as large as possible.
Design considerations:

1. When pad is at least 3X hole size, specify a min. .005 annular ring around the hole. (Figure 17-16.)
2. When pad is 2X hole size, specify that copper must be visible all around the hole and a .005 annular ring must cover 270° around the hole. (Figure 17-17 and Figure 17-18.)
3. Tie the pads down with ears when possible, but not at the cost of creating potential shorts. (Figure 17-19.)
4. Fillet all sharp bands to reduce stress concentrations. (Figure 17-20.)

To improve solderability and mechanical strength, use maximum copper area within circuit to get most copper around the hole. (See Figures 17-21, 17-22 and 17-23.)

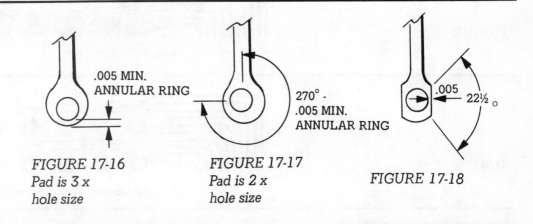

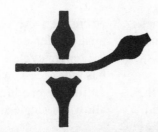

FIGURE 17-16
Pad is 3 x
hole size

FIGURE 17-17
Pad is 2 x
hole size

FIGURE 17-18

FIGURE 17-19

FIGURE 17-20

FIGURE 17-21

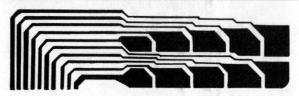

FIGURE 17-22

Maximum copper area

FIGURE 17-23

Relatively poor use of available circuit real estate.

ACCESS WINDOWS

Most circuits are insulated on both sides. Access windows must therefore be designed through one or both sides of the issulation. (As shown in Figure 17-24 and 17-25).

Insulation windows should overlap the pads as much as possible, providing additional strength for the pads. This way, they strengthen the circuit. Not just for the original solder connection but for as many a 10 repair cycles. The least expensive circuit has all access windows on the same side although product designs do not alway allow for single-sided access.

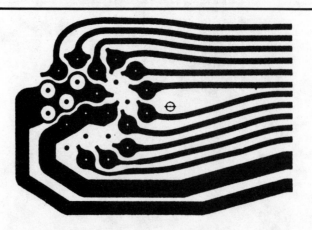

FIGURE 17-24
Insulation windows

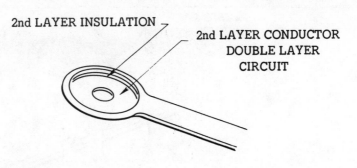

2nd LAYER INSULATION

2nd LAYER CONDUCTOR
DOUBLE LAYER
CIRCUIT

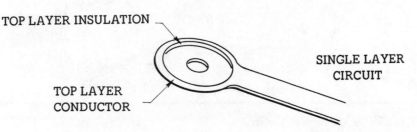

TOP LAYER INSULATION

SINGLE LAYER
CIRCUIT

TOP LAYER
CONDUCTOR

FIGURE 17-25
Access to single
and multilayer
circuits.

TERMINAL ACCESS WINDOWS

Three common techniques are used to remove or mask the cover coat from the terminals pads. The "individual" and "combination" are the most difficult to manufacture. The "strip technique" is gaining acceptance with the increased use of wave soldering equipment which now can be employed without bridging where the exposed copper areas are a minimum of 10 mils airgap. (See Figure 17-26 and 17-26A).

COMBINATION

INDIVIDUAL

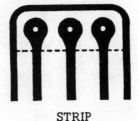

STRIP

FIGURE 17-26
Covercoat removal
above terminal pads.

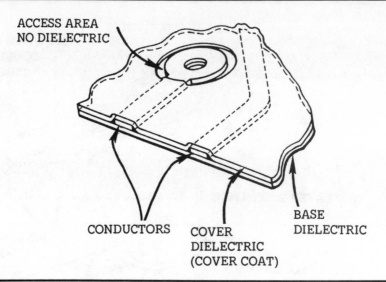

ACCESS AREA
NO DIELECTRIC

CONDUCTORS

COVER
DIELECTRIC
(COVER COAT)

BASE
DIELECTRIC

FIGURE 17-26A

REVERSE TERMINAL MASKING

To avoid access windows on both sides of flex circuit (exposing the terminal pad on both sides), we suggest using one of the following techniques (Figure 17-27). Techniques like this should be considered; the additional cost of material is far more than offset by the elimination of the labor and process involved in reverse masking.

Figure 17-28 shows one way of connect-

ing traces to a board or connector; this arrangement would require three separate layers of circuitry.

Figure 17-29 shows the best trace layout in a single layer. Thus, the basic principle of fanning out from the areas of greatest conductor density to areas of lesser density provides the best arrangement.

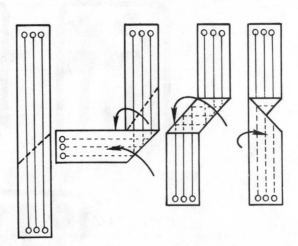

FIGURE 17-27

FLEXIBLE CIRCUITS

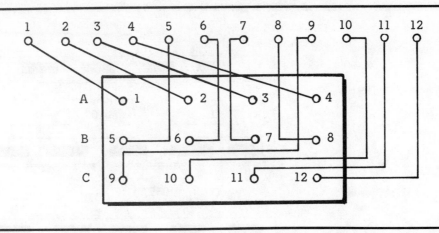

FIGURE 17-28

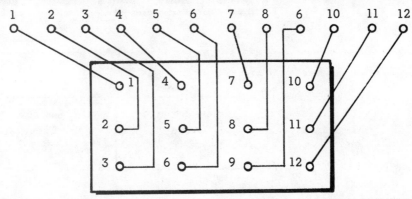

FIGURE 17-29

SHIELDING FLEX CIRCUIT

Shielding may affect mechanical flexibility of the circuit; as the shielding increases, flexibility usually decreases. (See Figures 17-30, 17-31 and 17-32.)

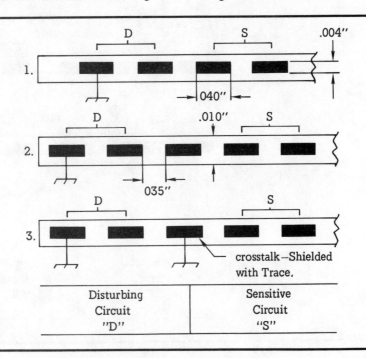

FIGURE 17-30

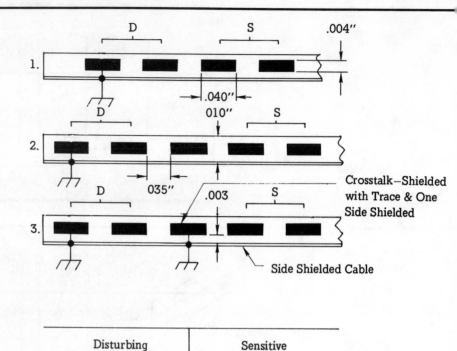

FIGURE 17-31

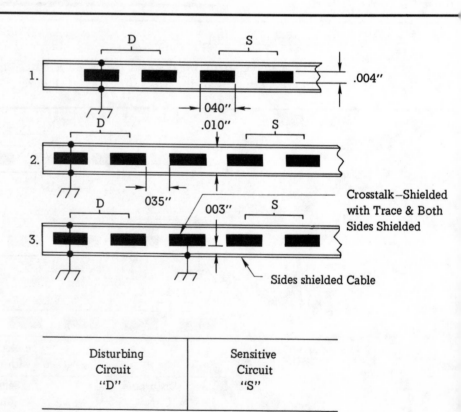

FIGURE 17-32

FLEXIBLE CIRCUITS

TOLERANCES

The following are general rules about tolerances that should be kept in mind while designing flex circuits. These recommended minimums are for economical production, regardless of lot size.

A- Center distance between holes

Drilled by eye – ± 0.010 inch
Punched with one single stage punch and die – ± 0.002 inch
Punched with two single stage punch and die – ± 0.005 inch

B- Hole to pattern tolerances

Drilled by eye pattern within 0.010 inch at center.
Punched with compound punch and die within 0.005 inch at center.
Registration front to back ± 0.010 inch.

C Hole diameter tolerance

Hole Size
Up to 1/16 inch diameter
5/64 to 3/16 inch diameter
Over 3/16 inch diameter

Hole Size- can be punched as small as .020 inch, but standard practice .030 inch or larger.

D- Hole to outside dimension tolerance

Piercing and blanking die – ± 0.010 inch
Drilling and shearing – ± 0.032 inch.

E- Circuit pattern to outside dimension

Sheared edges ± 0.032 inch
Blanked edges ± 0.010 inch

F- Over-all dimension tolerance

Sheared edges ± 0.032 inch
Blanked edges ± 0.010 inch

G- Etching tolerances for solder plate resist

Weight of copper per sq. ft.	Width Tolerance	Spacing Tolerance
1 oz.	±0.002 inch −0.002 inch	±0.002 inch −0.002 inch
2 oz.	±0.002 inch −0.003	±0.003 inch ∓∫∴
3 oz.	+0.002 inch −0.005 inch	+0.005 inch −0.005 inch

H- Plating of a minimum thickness only should be specified

MATERIALS

Foil—conductive foils are typ. electrodeposited or rolled copper, occasionally—aluminum & nickel/copper. Most use 1- or 2 oz. copper.

INSULATION

Considerations: Cost, dimensional stability, thermal properties and toughness.

$1.65/LB for Polyester Film to $25/LB for Polyimide Film.

POLYESTER FILM—100°C

NON-WOVEN DACRON EPOXY—150°C

NON-WOVEN POLYAMIDE EPOXY—150°C

TEFLON FILM—200°C

POLYAMIDE FILM—240°C

TFE TEFLON FILM—260°C

POLYIMIDE FILM—300°C

Most insulations are .001 or .002 in. thick up to .005

Chapter

18

EMI
(Electro-Magnetic
Interference)

This chapter introduces the art of shielding pc boards against EMI.

EMI stands for electro-magnetic interference, and sometimes it is called RFI for radio-frequency interference. Either way, it is harmful because it acts on sensitive electronic circuits in a way that is similar to direct sunlight acting on photographic film: It does unwanted, uncontrolled things to the object of our efforts. Hence it is "interference."

There are close similarities between EMI and visible light: Both are frequency spectrum propogations; both leak through cracks in an enclosure, and both can be filtered. Their main differences are that visible light is of a much higher frequency and is far more difficult to generate. It takes a conscious effort to generate visible light, but EMI we get for nothing. It is everywhere.

This is because at frequencies above 100 kHz or so, electronic signals just seem to leap out of the wires that were supposed to contain them. Ordinary wires become radiating antennas—and that is the fundamental principle of commercial radio and tv broadcasting today.

To get rid of EMI's effects, designers usually enclose the most-sensitive parts of a circuit in a metal box—the same as a light-proof box protects photographic film.

The past 15 years have given rise to a dramatic increase in the use of high-frequency technology in communications, instrumentation, medicine and computation. In many instances, equipment and system designers are unaware of the need and method of design considerations for electromagnetic compatibility. This design requirement has been brought about by the increased use of high-frequency techniques within equipment, and also by a vast increase in the number of users of the electromagnetic spectrum. Other changes, such as miniaturization through integrated circuit technology have forced equipment into small volumes requiring even better EM compatibility.

Thus the equipment designer now more than ever is faced with the problem of incorporating EMI suppression methods into his design. To apply these methods effectively, the designer must have a basic understanding of the methods. All too often these considerations have been ignored or not understood. This led to failure of devices and systems.

A HISTORY OF ELECTRO-MAGNETISM

The first knowledge of plain magnetism was obtained by the ancient Greeks who observed that certain varieties of iron ores had the ability to physically attract pieces of iron. To the best of our knowledge, the Greeks made no use of this knowledge during that period of time. Fourteen hundred years later (about 1000 A.D.) European navigators were using suspended magnets (the first form of magnetic compasses) for the guidance of their seagoing vessels.

About two centuries later, a man named

Peter Peregrinus shaped a piece of lodestone, a strongly magnetic iron ore, into a sphere. By placing a needle at various points on the sphere and drawing lines tangent to the needle, he found these lines converging, like meridians, towards two points on the sphere's surface. He called these points POLES.

The observations of Peregrinus were further enhanced by English physician named William Gilbert. He wrote a book entitled "De Magnete" in the year 1600 in which he claims he performed the same experiments with the lodestone sphere as Peregrinus did but, more importantly, claimed that it was a miniature counterpart to the earth itself—implying that the earth possessed the same magnetic characteristics. Subsequently, he floated two magnets in a fluid and demonstrated, for the first known time in history, that the poles of the magnets were different, and that like poles repel and unlike poles attract each other.

Gilbert was mainly known for his contributions in magnetism although he also contributed a great deal to the knowledge of the behavior of electrified bodies by demonstrating that bodies similar to amber have the ability, when charged through rubbing, to attract objects. He called these bodies ELECTRICS after the Greek word "elektron" meaning amber.

Later in the early 1700's a man named DuFay recognized and stated his belief in the existence of two kinds of electricity, later being called "+" and "−" by Benjamin Franklin. In the same period of time other men of science discovered that electricity from a charged sphere could be stored in a bottle if it contained copper wire or mercury. Work by others improved this apparatus by placing metal on the outside as well as in. This device was later to become known as the Leden jar and is similar to what we call a CAPACITOR today.

In 1752 Benjamin Franklin performed his famous kite experiment. He had noted that a metallic point will discharge an electrified conductor, and he proposed and executed his famous experiment to extract charge from a cloud. Needless to say, Franklin's work brought him permanent fame plus a lot of criticism from local pulpits for "interfering with the wrath of God."

Following Franklin's work, men like Coulomb and Volta contributed much to the understanding of electricity and charge. Coulomb attached quantity to scientific description of electric charge while Volta, in 1800, made the first battery or "voltaic pile" by placing discs of copper, zinc and moistened pasteboard in series. He discovered that upon closing a circuit around the battery, the arrangement gave a continuous electrical effect (current), whereas the Leyden jar had to be recharged after each short discharge. Hence, the first tangible source of current.

Following this period, in 1820, Oersted accidentally noticed the deflection of a compass placed near a conductor carrying an electric current and simultaneously discovered the converse: the deflection of an electric current by a magnet.

The early 1800's provided continued contributions to the world of electromagnetics with Faraday's invention of the first direct-current motor in 1821 and the construction of the first electromagnet by Sturgeon in 1825. Following this, in 1827, was the birth of the famous Ohm's Law $(E=IR)$. The results of Ohm's experiments relating to the presence and quantity of current along with Ampere's investigations of its magnetic and force effects during earlier years, form the foundations for the applications of electricity and magnetism to areas of engineering as they are today.

After this period in time, many other scientists made great contributions to the discovery and understanding of magnetic principles. Several of these many great men made notable discoveries. The first was Michael Faraday. In 1831 he discovered that a changing electric current induces an electromotive force in a properly placed conductor. This monumental discovery was not put to any use for over forty years.

The second was James Clerk Maxwell. In

1873, Maxwell compiled his famous electromagnetic theory that an electric field changing in time must be regarded as a "displacement electric current."

Incorporation of this idea into previous theory led to his world-famous Maxwell's equations, which mathematically state that electric and magnetic fields, if changing in time, generate each other, and by their combined action propagate energy as waves out into space at the velocity of light. It was not long after Maxwell's assertion that light was determined to be an electromagnetic wave phenomenon. The impact of Maxwell's work is massive.

The third man was Heinrich Hertz who, fifteen years after Maxwell's work, was able to verify the prediction as to the electromagnetic wave character of light by discharging an induction coil across a gap which emitted electromagnetic waves. These waves were received at the other end of Hertz's laboratory and caused sparks to jump across a gap placed in a metallic ring. By 1901, Guglielmo Marconi using these electromagnetic waves, was successfully transmitting messages by "wireless," our first form of radio. Thus, the work of James Clerk Maxwell will always remain one of the greatest contributions to science.

And today, Section 302 of the Federal Communication Act makes the manufacturer, as well as the user, responsible for electronic interference caused by equipment. The Federal Communications Commission regulates use of radio and wire communications, which includes all interference caused by commercial, public, or private equipment as well as interference with such apparatus. Military electronic hardware is regulated by requiring compliance with MIL-STD-461. (See Figure 18-1.)

In its broadest definition, EMI (Electro-Magnetic Interference) is any electromagnetic energy that causes an undesirable response. In most cases, EMI causes a malfunction to occur in other electrical or electronic equipment. However, it is possible that the source of EMI and the malfunction due to it can occur in the same equipment or system.

"Keep it in or keep it out"—there are four basic types of EMI:

1. System Radiated EMI
2. External Radiated EMI
3. Internal Radiated EMI
4. Conducted EMI

(See Figure 18-2.)

The block diagram Figure 18-1 shows the categories of test specified by MIL-STD-461. Note that tests are required for both conducted and radiated noise signals. Equipment designed to comply with MIL-STD-461 must pass testing and evaluation as shown above.

The importance of knowledge and early consideration of EMC (electromagnetic compatibility) in the design and development of electronic devices and systems is shown by Figure 18-3. The onset of an equipment design is clearly the time at which a designer has the maximum number of EMC design techniques available for implementation; the minimum number of constraints will exist on the design at this time.

EMI CONTROL There are three basic ways of controlling EMI:

1. Redesign the source so that it *produces* less EMI.
2. Redesign the malfunctioning equipment so that it is less likely to *respond* to EMI.
3. Initially design for electromagnetic compatibility.

There are two generally accepted methods for containing or excluding EMI: filtering and shielding.

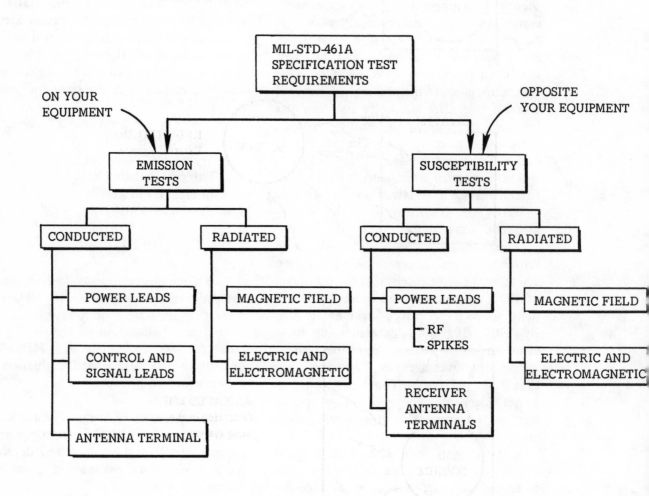

FIGURE 18-1
Required test for
both conducted and
radiated noise
signals per
MIL-STD-461

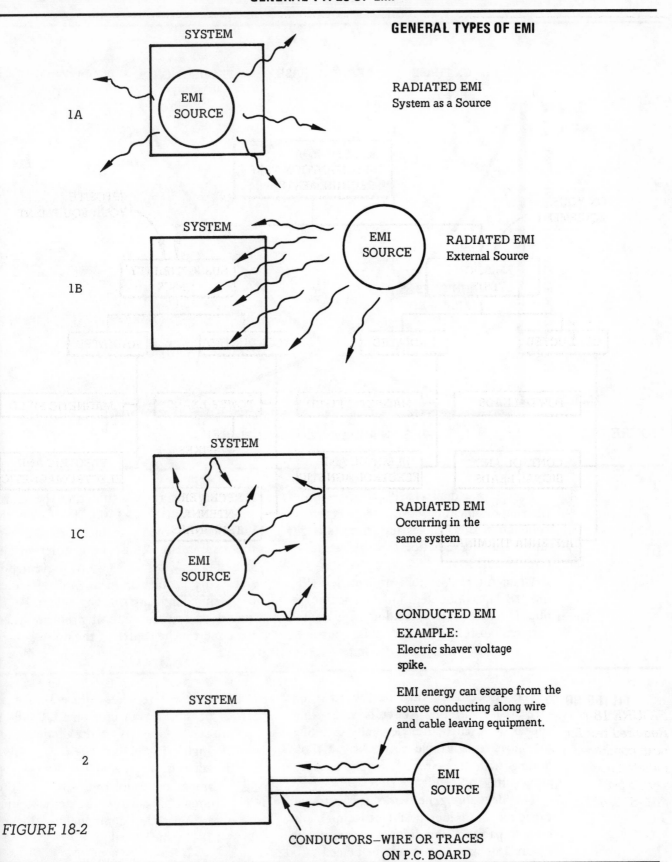

GENERAL TYPES OF EMI

1A

RADIATED EMI
System as a Source

1B

RADIATED EMI
External Source

1C

RADIATED EMI
Occurring in the
same system

CONDUCTED EMI
EXAMPLE:
Electric shaver voltage
spike.

EMI energy can escape from the
source conducting along wire
and cable leaving equipment.

2

FIGURE 18-2

CONDUCTORS—WIRE OR TRACES
ON P.C. BOARD

EMI (ELECTRO-MAGNETIC INTERFERENCE)

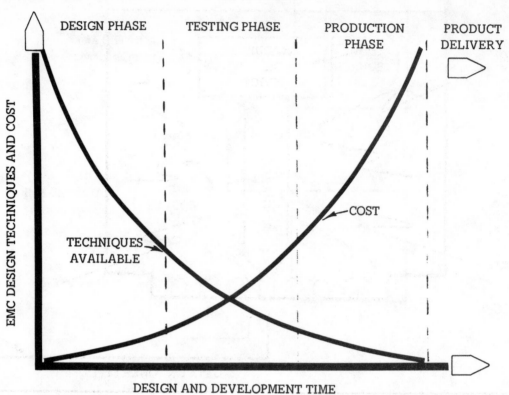

FIGURE 18-3

Figure axis labels: EMC DESIGN TECHNIQUES AND COST (vertical), DESIGN AND DEVELOPMENT TIME (horizontal), DESIGN PHASE, TESTING PHASE, PRODUCTION PHASE, PRODUCT DELIVERY, COST, TECHNIQUES AVAILABLE

Figure 18-4 depicts a PC Board without filtering or shielding. EMI energy enters by conduction along cables and traces (straight arrows) and by radiation throughout the air (crooked arrows).

When filter capacitors are installed (Figure 18-5), conducted interference is reduced. These filters conduct the desired currents but reject the undesirable currents.

However, energy can still enter the sourc by direct radiation. A shield prevents th radiated EM energy from interfering wit the equipment. Thus, both filtering an shielding may be necessary to prevent po sible EMI problems. In a similar manne both shielding and filtering might be r quired to contain an EMI problem when the circuit is the source of the noise.

FILTERING

The PC designer must be able to recognize basic "filtering" circuits when they appear on a schematic. Generally speaking, the most recognizable networks are those dealing with the circuits' supply or operating voltages. See Figure 18-6.

The decoupling networks (resistor and capacitor or inductor and capacitor), discrete capacitors and bypass capacitors in Figure 18-6 and 18-7 should be placed:

a) as close to the voltage source a possible (connector pins 1, 2, and or the outputs of the diode bridg Figure 18-6A thru 18-6D.

b) adjacent to the active devices (op amps, or transistors, etc.) of th circuit. Resistors, capacitors, etc are passive components, Figure 18-7A and 18-7B.

EMI (ELECTRO-MAGNETIC INTERFERENCE)

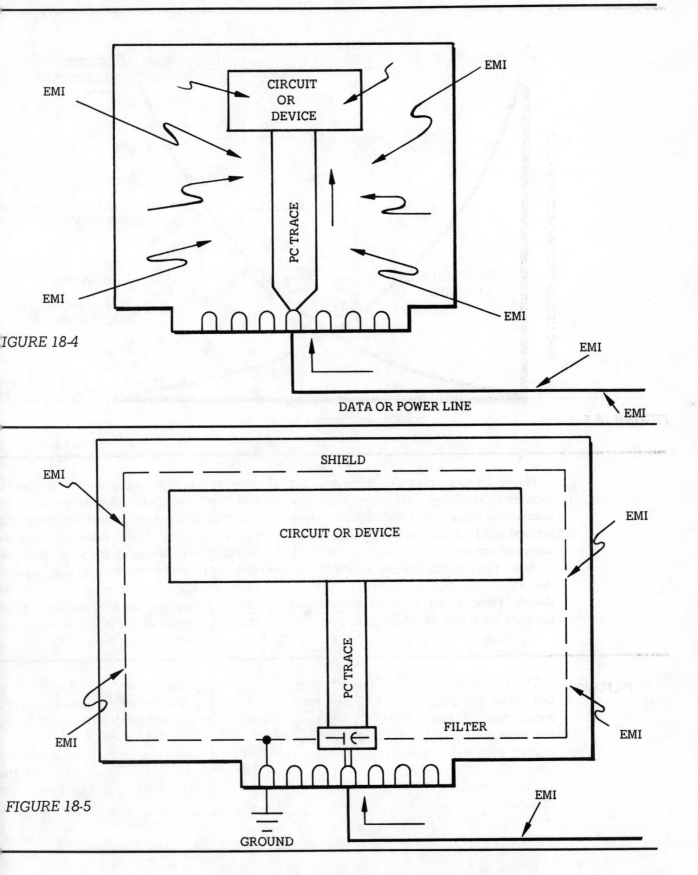

EMI

EMI

CIRCUIT
OR
DEVICE

PC TRACE

EMI

EMI

FIGURE 18-4

EMI

DATA OR POWER LINE

EMI

EMI

SHIELD

EMI

EMI

CIRCUIT OR DEVICE

PC TRACE

FILTER

EMI

EMI

FIGURE 18-5

GROUND

EMI

EMI

EMI (ELECTRO-MAGNETIC INTERFERENCE)

301

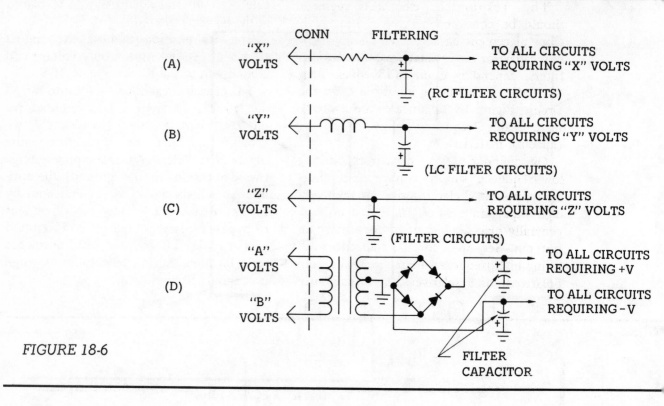

FIGURE 18-6

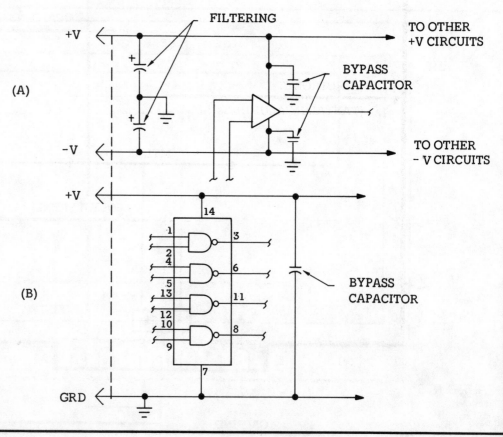

FIGURE 18-7

The responsible electrical engineer should be consulted as to the most judicious placement on an analog circuit. Various capacitor types (materials) are used as filters, depending upon the frequency of the EMI "noise" being impinged upon the circuit. Figure 18-8 delineates the approximate usable frequency range for various capacitor materials.

On digital circuitry, capacitors filtering low-frequency supply voltage noise (normally, polarized electrolytics of a relatively high capacitance value $10\mu F$ and up) are generally placed adjacent to the voltage input pin. See 18-6 and -7. Capacitors filtering high frequency noise (normally non-polarized, of a relatively low capacitance value— .1 μF and down) should be placed at the IC being filtered.

The ideal position for a capacitor and its IC would be across the supply voltage and ground pins of the IC. See Figure 18-9.

The capacitor cannot be built into the IC due to the different values required for different circuit uses. Nor should we "stack" components as shown. In that the object is to "clean-up" the supply voltage (we're not trying to filter ground), the optimum positions would be as indicated by Figure 18-10A and B. However, placement of bypass capacitors adjacent to IC ground pins may be, under some conditions, the ideal location. Again, the responsible engineer should be consulted.

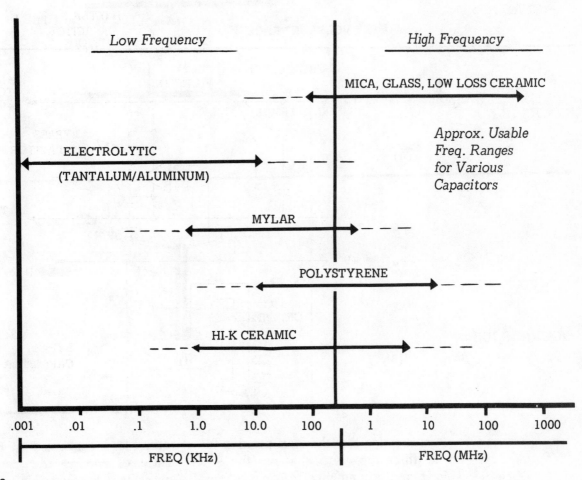

FIGURE 18-8

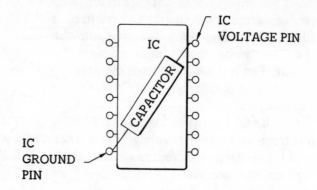

FIGURE 18-9

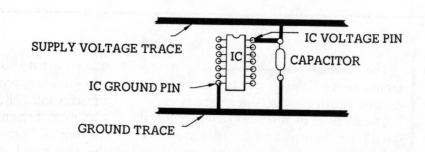

FIGURE 18-10A

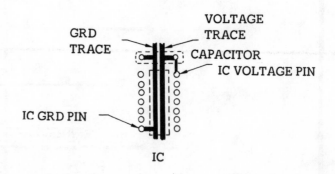

FIGURE 18-10B

Remembering that the ideal position for an IC filter capacitor is across the IC's voltage and ground pins, capacitors must *never* be placed with one lead connected to the Voltage bus trace of one row and the other lead connected to the ground b[us] trace of another row. Such a capacit[or] placement has no value to the circuit. S[ee] Figure 18-11.

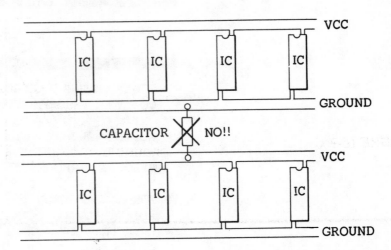

FIGURE 18-11

The voltage and ground bus traces should be as close to the IC's as possible, consitent with good PC design practices.

The number of filter capacitors required should be discussed with the responsible electrical engineer. If the requirements are one capacitor for every three IC's and there are thirteen IC's, you may need to place five capacitors on the board. However, too many capacitors may cause other electrical problems (capacitive inductance). See Figure 12A and 12B.

This is presently being solved through the use of "capacitor bus bars" in which the two metal strips (one for power and one for ground) themselves act as a capacitor. See Figure 18-13, 18-14A and B.

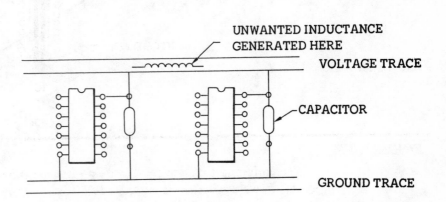

FIGURE 18-12A

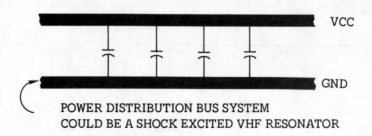

POWER DISTRIBUTION BUS SYSTEM
COULD BE A SHOCK EXCITED VHF RESONATOR

- ADDITIONAL DECOUPLING CAPS
  WILL TEND TO SHIFT THE
  RESONANT FREQUENCY

- POSSIBLE SOLUTION: USE AN
  ADDITIONAL LOSSY CHOKE AND
  CAPACITOR FOR EACH (OR PAIR)
  OF IC'S

FIGURE 18-12B

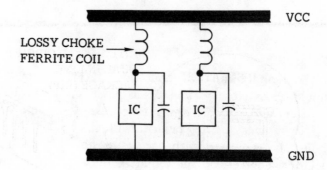

FIGURE 18-13

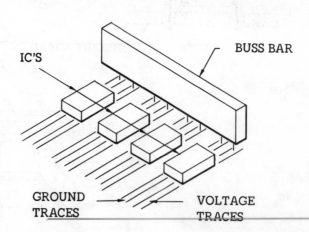

EMI (ELECTRO-MAGNETIC INTERFERENCE)

Bus bars can provide

- Greater component/wiring densities without going to multilayer boards

- Reduce power distribution inductance and resistance

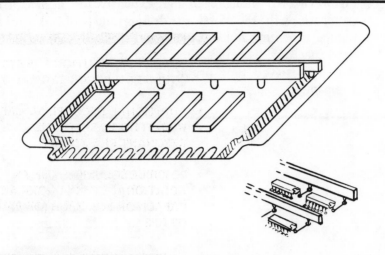

FIGURE 18-14A

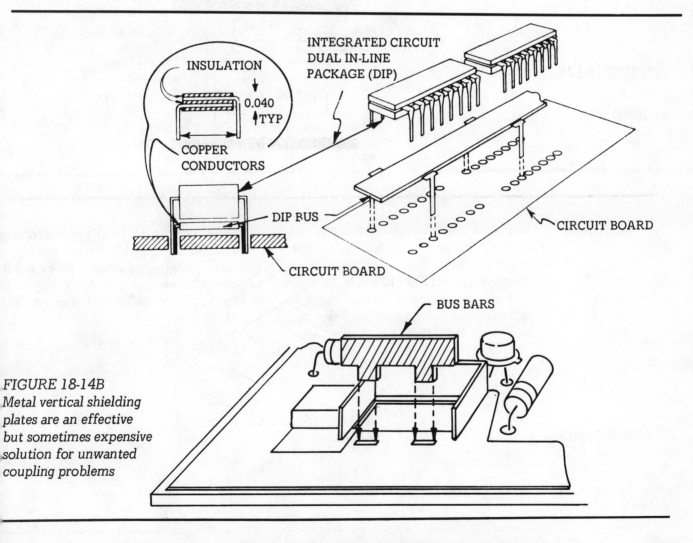

INSULATION

0.040 ↑ TYP

COPPER CONDUCTORS

INTEGRATED CIRCUIT DUAL IN-LINE PACKAGE (DIP)

DIP BUS

CIRCUIT BOARD

CIRCUIT BOARD

BUS BARS

*FIGURE 18-14B
Metal vertical shielding
plates are an effective
but sometimes expensive
solution for unwanted
coupling problems*

EMI (ELECTRO-MAGNETIC INTERFERENCE)

A number of companies are presently designing into their PC boards internal voltage and ground planes, spaced a specified distance apart, they act as a capacitor. (See Figure 18-14C) Though seemingly expensive (compare with the costs of drilling and plating mounting holes, buying and stocking capacitors, assembly time, etc.) it has definitely solved EMI problems for these companies.

EMI filter networks (inductors and capacitors), encapsulated for mounting directly to a PC board, are available from several manufacturers. These firms specialize in the design of filters for all kinds of conducted EMI problems, and are listed in standard industrial product directories.

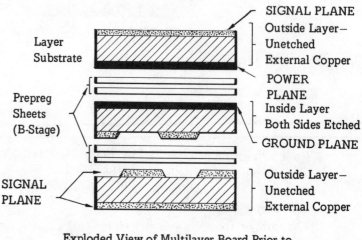

FIGURE 18-14C

Exploded View of Multilayer Board Prior to Lamination Step (6 Copper Layers)

SHIELDING Control of EMI through shielding is not as easily solved by the designer as the use of proper filtering. Some knowledge of the electromagnetic field is necessary. Simple physics can demonstrate that two energy fields, (electrical & magnetic) are generated when a wire (PCB trace) is charged with electrons. (Figure 18-15A) The electromagnetic field, with which we are dealing, is not static because its intensity varies and its polarity alternates. The rate at which the field alternates is called its frequency (Figure 18-15B).

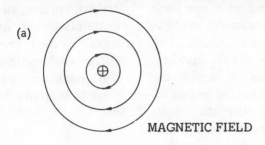

FIGURE 18-15A

MAGNETIC FIELD

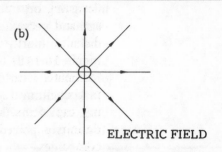

ELECTRIC FIELD

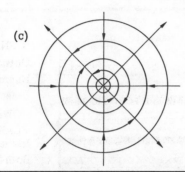

FIGURE 18-15B

ALTERNATING CURRENT

Four electromagnetic field characteristics are important in shielding theory: *a. Frequency b. Wavelength c. Strength (amplitude) d. Impedence.*

The unit of frequency is the Hertz (Figure 18-16) and it equals one cycle in one second. Frequencies up to several billions of cycles per second are normal in EMI work.

Wavelength of an electromagnetic field is the distance traveled in one full cycle at the speed of light. For instance at 1KHZ, the wavelength is the distance traveled in 1/1000 of a second (186 miles).

Strength (amplitude) is a measure of the field's power in either so many watts/meter² or volts/meter or amps/meter. We can, using proper instruments, measure the fields.

Impedence compares these intensities in the following manner:

$$Impedence = \frac{electrical\ field\ intensity\ (strength)}{magnetic\ field\ intensity\ (strength)}$$

Of the above four, only a shield has an effect on the intensity. The other three characteristics effect "the shield" and "its use."

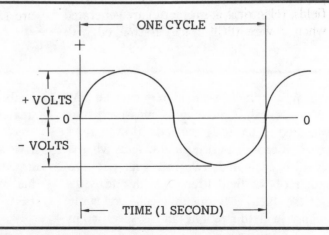

FIGURE 18-16

1 HERTZ (HZ) EQUALS
1 CYCLE PER SECOND

Shielding is accomplished by two mechanisms: Absorption loss and reflection loss. At low frequencies, shielding is accomplished primarily by absorption loss, and at high frequencies shielding is accomplished primarily by reflection loss. However, the total shielding function is a composite of both reflection loss and absorption loss. These two phenomenae have been simpli-

fied for calculating the thickness of a shield. Science has defined a unit of measurement for shielding called "one skin depth." This is the thickness of a piece of metal which if bombarded with radiated EMI would reduce the EMI signal by 37% of its original intensity after passing through the metal. (See Figure 18-17A and -17B).

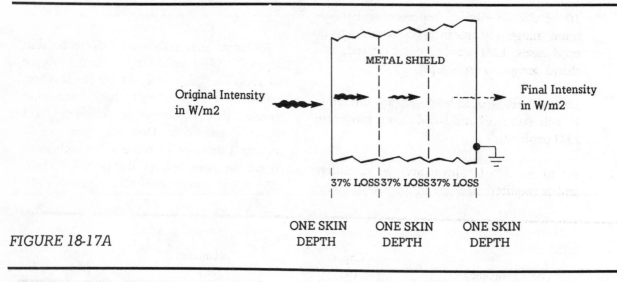

FIGURE 18-17A

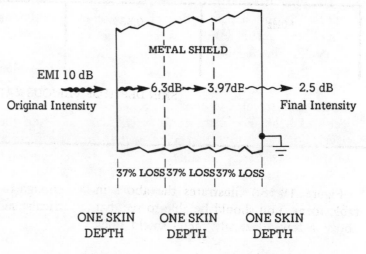

FIGURE 18-17B

Simply stated, a shield reduces the EM field strength. Shielding is measured and specified in terms of the reduction in field strength caused by the shield. Thus when a shield is evaluated, two measurements are required: the field strength with and without the shield. The shielding achieved is the change in field strength, and is specified in

decibels. Since the decibel is the universal measure for shielding, it is very important to clearly understand its meaning and to have a "feel" for its magnitude. A decibel measures the reduction in field strength due to a shield, and not the field strength itself.

EMI (ELECTRO-MAGNETIC INTERFERENCE)

The following descriptions will impart some feel for what 30 dB or 50 dB or even 100 dB means.

0 to 10 dB—is very little shielding. An enclosure that reduces an EM field by this amount hardly deserves to be called a shield. EMI, generally speaking, would not be eliminated.

10 to 30 dB—Would represent the minimum range for meaningful shielding. In mild cases, EMI would be eliminated, and shield design is very simple.

30 to 60 dB—Would be "average" shielding. It will solve all mild and some moderate EMI problems.

60 to 90 dB—Is above average shielding, and is required to solve moderate to severe EMI problems. Shield design is of primary importance in equipment-housing design.

90 to 120 dB—Generally speaking, is the maximum possible with the best shielding designs. Measurements require instrumentation specifically designed for this purpose, and in some cases, measurement is beyond present state-of-the-art.

Different materials have different "skin depth." The following chart (Figure 18-18A) illustrates this. At 60 Hertz, there would be a 37% intensity loss if one were to use a sheet of copper approximately 1/3 of an inch thick. However, you would require a sheet of aluminum 1/2 inch thick to get the same results. But only 1/32 inch steel would absorb an equal amount of EMI.

Frequency	Copper (in.)	Aluminum (in.)	Steel (in.)
60Hz	0.335	0.429	0.034

FIGURE 18-18A   SKIN DEPTH OF VARIOUS MATERIALS

Figure 18-18B illustrates the above in table form. You should be able to see that above a few MHz, any metal shield thick enough to be mechanically suitable is electrically suitable.

FREQUENCY	COPPER (IN.)	ALUMINUM (IN.)	STEEL (IN.)
60 Hz	0.335	0.429	0.034
100 Hz	0.260	0.333	0.026
1 KHz	0.082	0.105	0.008
10 KHz	0.026	0.033	0.003
100 KHz	0.008	0.011	0.0008
1 MHz	0.003	0.003	0.0003
10 MHz	0.0008	0.001	0.00008

FIGURE 18-18B
Skin Depth of
Various Materials

Note that the amount of shielding due to reflection has not yet been considered. Shield *reflection* is not as easy to explain nor calculate as shield *absorption*. Some of the energy is reflected at the surface due to the shield's inability to accept all the energy. The impedance (one of the four previously mentioned characteristics) of the field determines how much will be reflected. A high impedance field (also called an "electrical field") will have more reflec-

tion than a low impedance field (also known as a "magnetic field").

A shield contains (or excludes) electro-magnetic energy by reflecting or absorbing the energy. Whenever EM passes from one medium into another, a portion is reflected. Some energy does penetrate into the shield; the amount absorbed by the shield depends on the field's frequency and the shielding materials and thickness.

EMI (ELECTRO-MAGNETIC INTERFERENCE)

SHIELDING MATERIAL CHARACTERISTICS

Relative CONDUCTIVITY and PERMEABILITY of
VARIOUS MATERIALS

MATERIAL	Relative CONDUCTIVITY (high Freq.)	Relative PERMEABILITY (Low Freq.)
Silver	1.05	1
Copper-Annealed	1.00	1
Gold	0.70	1
Aluminum	0.61	1
Brass	0.26	1
Nickel	0.20	1
Bronze	0.15	1
Steel-SAE 1045	0.10	1000
Lead	0.08	1
Monel	0.04	1
Stainless Steel (430)	0.02	500

FIGURE 18-19

ABSORPTION PLUS REFLECTION

Figure 18-20 shows total shielding of magnetic fields with a source-to-shield distance of 12 inches for copper, aluminum and iron, both 1/32 inch and 1/8 inch thick. Total shielding for all plane wave and electric fields would be more than 120 dB and could not be shown on this curve.

For low-frequency magnetic field shielding, reflection is so small, and not affected by shield thickness that total shielding can be increased only by increasing the EM energy absorption. Thus, low-frequency magnetic shielding is increased by;

1. The shield thickness
2. The shield conductivity

It can be deduced from the above that generation of a magnetic field on our PC board would be detrimental to the circuit. A loop antenna connected to a generator with a high RF current output generates an intense electro-magnetic field. If such a field were generated at a relatively low frequency, the results would be devastating.

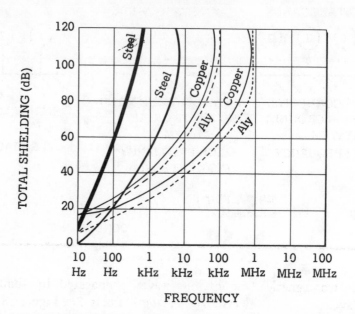

FIGURE 18-20

GROUNDING NOTES

A. Grounding is one of the primary ways to minimize unwanted noise and undesired signals.

B. A good ground system must be DESIGNED just like the rest of the devices and circuits comprising an electronic system.

C. THERE ARE THREE BASIC OBJECTIVES INVOLVED IN THE DESIGN OF A GOOD GROUNDING SYSTEM:

(1) Minimize the noise voltage generated by currents from two or more circuits, which are flowing through a common conductor.

(2) Avoid creating GROUND LOOPS and differences in ground potential.

(3) Keep signal traces short and direct to avoid their acting as receiving antennas.

CLASSES OF GROUNDING METHODS

A. At LOW frequencies, typically below 1 MHz, a SINGLE POINT ground system should be used.

B. At HIGH frequencies, typically above 1 MHz, a multipoint ground system should be used.

EMI (ELECTRO-MAGNETIC INTERFERENCE)

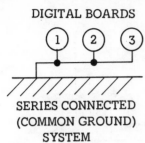

DIGITAL BOARDS

SERIES CONNECTED
(COMMON GROUND)
SYSTEM
LOW FREQUENCY

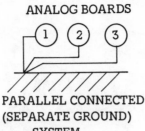

ANALOG BOARDS

PARALLEL CONNECTED
(SEPARATE GROUND)
SYSTEM
HIGH FREQUENCY
(NOT THE BEST)

SINGLE-POINT
GROUND

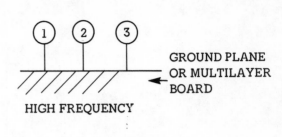

GROUND PLANE
OR MULTILAYER
BOARD

HIGH FREQUENCY

MULTIPOINT
GROUND

FIGURE 18-21
Classes of grounding
methods

DEFINITIONS AND TECHNIQUES		
Electromagnetic Wave:	One of the waves propagated by simultaneous periodic variations of electric and magnetic fields. See Figure 18-22;
EMI:	Electromagnetic Interference. See Figure 18-23.
Electromagnetic Shield:	A partition, can or box made of magnetic (ferrous) material (iron, steel, or special alloy) enclosing a magnetic component. See Figure 18-28.
EMI:	Electromagnetic interference—disturbances of equipment operation caused by electromagnetic fields from outside/inside sources. See Figure 18-23.
EMI Gasket:	A material used for sealing seams and joints on a shield, to prevent EMI energy from passing through it. See Figure 18-24.
Frequency (Hertz):	The number of complete cycles per second of an alternating current. See Frequency Spectrum Chart. Figure 18-25A and B.
Impedance:	The total opposition offered by a circuit or device to the flow of alternating current. See Figure 18-26.
Interference:	The disturbing effect of any undesired signal. See Figure 18-27.
Leakage:	When EMI energy passes or enters through a shield (seams, joints and openings). See Figure 18-28, 29 and 30.
Line Filter:	A circuit or device used for suppressing or minimizing certain unwanted frequencies from signal or power conductors. See Figure 18-32, 33.
Noise:	A signal or a composite of signals appearing in a circuit other than the desired signals.
Radiated EMI:	Electromagnetic waves traveling in air. See Figure 18-23.
RFI:	Radio frequency interference.
Shield:	A material used to suppress EMI energy to a tolerable limit. See Figure 18-28.
Susceptibility:	The capability of a device or circuit to respond to unwanted electrical energy (noise).

EMI (ELECTRO-MAGNETIC INTERFERENCE)

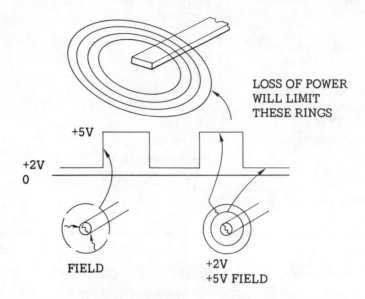

LOSS OF POWER
WILL LIMIT
THESE RINGS

+5V

+2V
0

FIGURE 18-22
Electromagnetic wave

FIELD

+2V
+5V FIELD

(EMI) ELECTROMAGNETIC INTERFERENCE

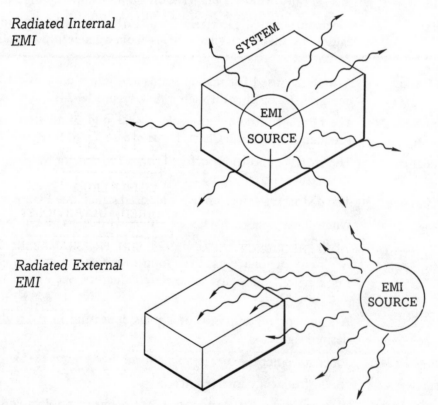

*Radiated Internal
EMI*

SYSTEM

EMI
SOURCE

*Radiated External
EMI*

EMI
SOURCE

FIGURE 18-23

EMI (ELECTRO-MAGNETIC INTERFERENCE)

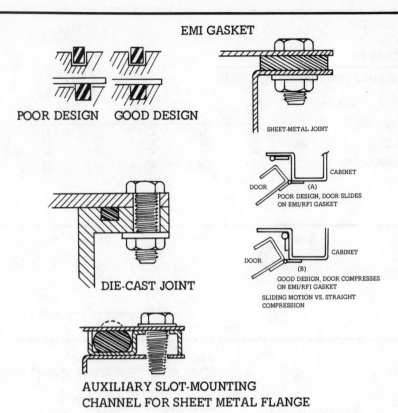

EMI GASKET

POOR DESIGN   GOOD DESIGN

SHEET-METAL JOINT

CABINET

DOOR
(A)
POOR DESIGN, DOOR SLIDES
ON EMI/RFI GASKET

CABINET

DOOR
(B)
GOOD DESIGN, DOOR COMPRESSES
ON EMI/RFI GASKET

SLIDING MOTION VS. STRAIGHT
COMPRESSION

DIE-CAST JOINT

AUXILIARY SLOT-MOUNTING
CHANNEL FOR SHEET METAL FLANGE

FIGURE 18-24

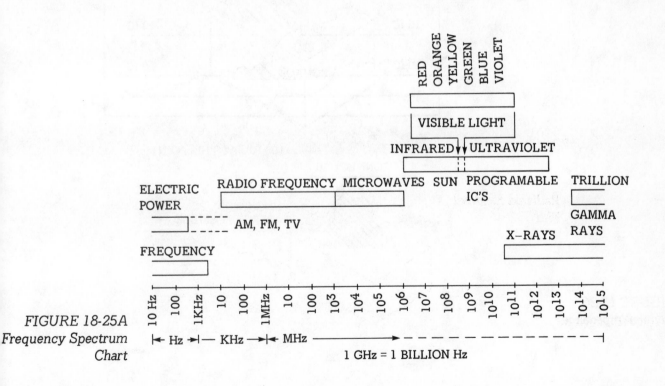

FIGURE 18-25A
Frequency Spectrum
Chart

EMI (ELECTRO-MAGNETIC INTERFERENCE)

(EXPANDED RADIO FREQUENCY PORTION)

		SHORT WAVES			MICRO WAVES

VLF	LF	MF	HF	VHF	UHF	SHF

3 KHz	30 KHz	300 KHz	3 MHz	30 MHz	300 MHz	3 GHz (3000 MHz)	30 GHz

FIGURE 18-25B
Frequency Spectrum
Chart

FREQUENCY

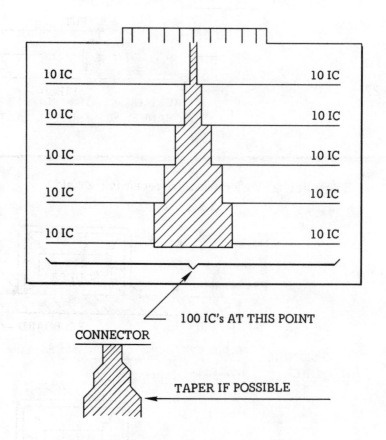

100 IC's AT THIS POINT

CONNECTOR

TAPER IF POSSIBLE

FIGURE 18-26
Reduce impedance

EMI (ELECTRO-MAGNETIC INTERFERENCE)

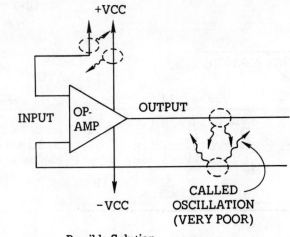

OUTPUT

INPUT

CALLED
OSCILLATION
(VERY POOR)

Possible Solution

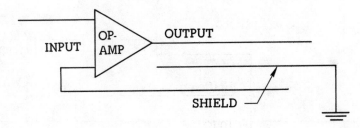

OUTPUT

INPUT

SHIELD

FIGURE 18-27
Interference

Solder Joints: *Very difficult to repair in the field*

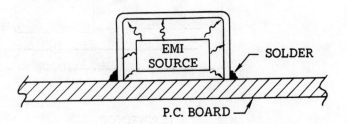

Bolting: *Sometimes, just bolting is enough*

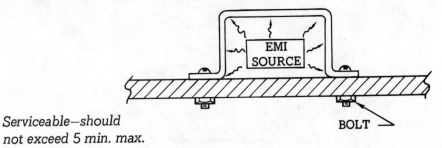

FIGURE 18-28
Leakage

Serviceable—should
not exceed 5 min. max.

EMI (ELECTRO-MAGNETIC INTERFERENCE)

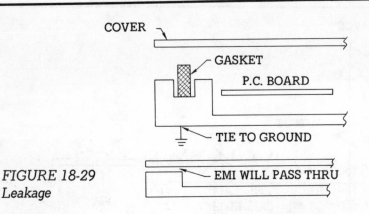

FIGURE 18-29
Leakage

MOST COMMON METHOD IS
GASKETING –

REAL-WORLD COMPLICATIONS–
Solid, continuous shields are seldom used in actual applications. Practical shields have discontinuities at covers, doors, panels, ventilating openings and panel hardware. Experience shows that these cracks and openings are often very leaky. In fact, EMI leakage above 1 MHz is due to discontinuities.

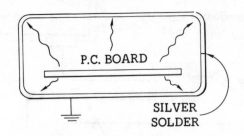

FIGURE 18-30
Leakage

WAVE LENGTH AFFECTS
HOW MANY SCREWS TO
USE, OR SIZE OF OPENING
LESS THAN
¼ WAVE LENGTH

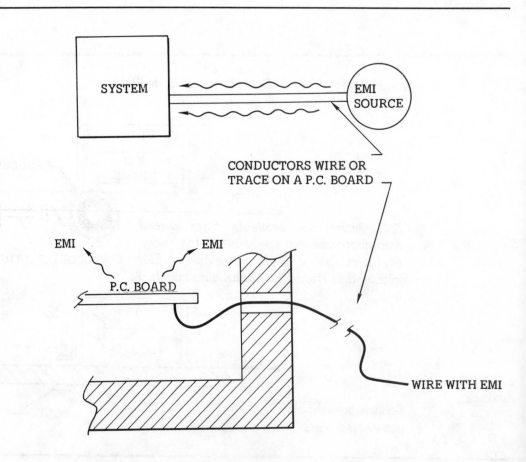

FIGURE 18-31
Conducted EMI

EMI (ELECTRO-MAGNETIC INTERFERENCE)

Example opposite shows how filtering and shielding combine to exclude EMI from susceptible equipment. If only filtering is installed, interference could still enter by direct radiation; if only shielding is used, interference would be conducted by the wiring through the shield and into the equipment.

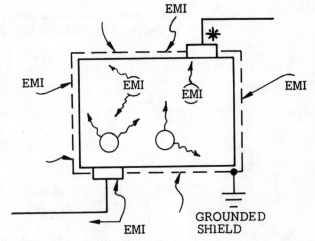

GROUNDED SHIELD

EMI filters are available from several manufacturers that specialize in the design of filters for all kinds of conducted EMI problems. They are listed in standard industrial product directories.

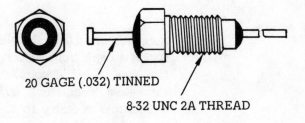

20 GAGE (.032) TINNED

8-32 UNC 2A THREAD

10MHZ TO 10 GHZ DC FILTERS

FIGURE 18-32
Filters

EMI (ELECTRO-MAGNETIC INTERFERENCE)

Bolted Feed-Through EMI Filters minimize
conducted pulse noise within a frequency
spectrum of 10 MHz to 200 MHz.

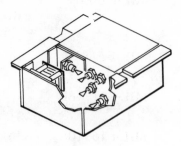

Microminiature Bulkhead Feed-Through
Filters provide low capacitive loading and
high insertion loss in microwave. Filters
readily pass dc control and power signals
and logic pulses but prevent module-to-
module microwave conduction.

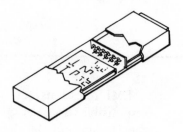

Subminiature UHF Amplifier Filter with
socket accepts transistor or thin film amplifier pins. Filter decouples dc power and
suppresses conducted UHF interference
from other circuitry to improve amplifier
performance. Filter is flow solderable.

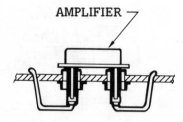

AMPLIFIER

FIGURE 18-33

MILITARY SPECIFICA-TIONS FOR EMI		
	MSFC-SPEC 279	Electromagnetic Compatibility (15kHz to 1-GHz).
	MIL-STD-826	Electromagnetic Interference Test Requirements and Test Methods (14kHz to 10Ghz).
	MIL-E-6051C	Electrical-Electronic System Compatibility and Interference Control Requirements for Aeronautical Weapons Systems, Associated Subsystems, and Aircraft (System Frequencies).
	MIL-1-6181D	Interference Control Requirements, Aircraft Equipment (15KHz to 10GHz).
	MIL-I-11748B	Interference Reduction for Electrical and Electronic Equipment (14kHz to 36GHz).
	MIL-I-16910C	Interference Measurements Radio Methods and Limits (14kHz to 10GHz).
	MIL-I-26600	Interference Control Requirements Aeronautical Equipments (15kHz to 10GHz).

NOTE: These specifications and standards may be obtained from:

COMMANDING OFFICER
U.S. NAVAL PUBLICATIONS & FORMS CENTER
5801 TABOR AVENUE
PHILADELPHIA, PENNA. 19120

EMI questions to ask yourself and the responsible engineer prior to the start of the Design and Development of your electronic device or system:

1. Each Designer should ask himself why a working knowledge and awareness of EMI Design consideration is important to a P.C. Designer.

EXAMPLE:

a. Communicating with the Engineer.
b. To ask applicable questions.
c. Choose correct material.
d. Able to see possible problems.
e. Know what type of problems arise from high frequency versus low frequency for shielding or filtering.
f. P.C. Board Design voltage or ground traces, thickness, ground plane, short connections & ground connections.

2. At what stage in an equipment design should EMI Design considerations be applied?
   ANSWER: Beginning of Design.
3. Referencing your answer to Question #2, *WHY*?
   ANSWER: At the beginning stage the cost of change is much lower than after the design is underway or complete.
4. The Decibel is a unit used to express electrical voltage, current and power_____? ANSWER: ratio
5. What government agency has primary responsibility for regulation of interference with and usage of wire and radio services?
   ANSWER: FCC
6. The basic unit of frequency is the _____? ANSWER: Hertz
7. *Radiated* electromagnetic fields are not created by conductors carrying *only*?
   ANSWER: Pure direct current.

8. All EMI Design problems are treated differently, based on frequency range and magnitude of the problem.

9. An interference problem in the form of *electromagnetic waves* is generally referred to as a *RADIATED* problem.

10. Conducted interference cannot occur at radio frequencies.
    ANSWER: FALSE

11. Shielding *effectiveness* is usually expressed in terms of:
    ANSWER: DECIBELS

12. The two primary mechanisms of a shield are:
    ANSWER: Absorption and reflection.

13. Skin effect is:
    ANSWER: 1. Frequency sensitive
    2. Material sensitive

14. The distance into a medium (shield material) required to reduce a field to 37% of its original value is called:
    ANSWER: Skin Depth

15. Generally speaking, *shielding* is:
    ANSWER: A composite effect of absorption and reflection.

16. A highly accepted method of *sealing discontinuities* in shield structures is:
    ANSWER: RFI gasketing

17. *Conducted* interference is usually suppressed by:
    ANSWER: RFI filters

18. If you were designing a ground system in which noise problems were *not* serious and simplicity was a must, you should use a:
    ANSWER: Common ground system

19. The ground system most likely t[o] be used for optimum performanc[e] at *Low Frequencies* is:
    ANSWER: Separate ground sy[s]tem

20. The ground system mainly intende[d] for *High Frequency* use:
    ANSWER: Multipoint groun[d] system

21. Regardless of the type of groun[d] system used in a system design yo[u] should always:
    ANSWER: Optimize ground im[-]pedance.

22. Once a material has been selecte[d] for a shielding job, the material i[t]self becomes a *lesser* concern tha[n]
    ANSWER: Leakage throug[h] seams and joints.

23. Which material would you specif[y] for a *Low Frequency* shielding job[?]
    ANSWER: Steel

24. What are the two main methods [of] EMI suppression we have studie[d]
    ANSWER: 1. Shielding
    2. Filtering

25. A good shielding design will:
    ANSWER: Reduce EMI to a to[l]erable level, but not to zero inte[n]sity.

26. A good high frequency P.C. Boar[d] layout should have:
    ANSWER: Very short leads an[d] traces.

27. Prior to starting a high-frequenc[y] layout a designer should:
    1. Review all critical areas of th[e] design with the responsib[le] engineer.
    2. Ask responsible enginee[r] about any possible RFI/EM[I] problems.

Chapter

19

CADD
(Computer Aided
Design
and Drafting)

This chapter introduces CADD—computer aided drafting and design systems, or CAD for short.

As the name implies, CAD systems are computer aids for the pc board designer. They are tools in the same sense as templates, straight edges and eraser shields. Being tools, CAD systems do not replace pc board designers, any more than automobiles replace travelers.

Instead of drawing a pc board layout on paper with pen and ink, the pc board designer with a CAD system at his disposal draws on a tv tube with a light pen. Drawings so produced are called "soft" (changeable), as opposed to "hard" (fixed). Software is changeable; hardware is not.

Since only the *information* representing the drawing is recorded, no hard evidence of it remains such as graphite, tape strips or ink. Subsequent alterations to the drawing are quick and easy since there is nothing to erase, lift or throw away.

CAD systems' greatest benefit to designers are their software programs. CAD programs allow pc board designers to determine the optimum routings for all traces on a board—simultaneously and at once. CAD systems are very powerful tools.

CAD systems are sometimes used remotely, by telephone lines. In these systems, many users share the cost of the hardware and software; they share time on the system. This is called timesharing.

Computers aid the product designer by performing the tedious calculations necessary to solve large design problems. Recent developments in computer technology have provided the designer with "Electronic Drawing Boards" in the form of video displays on which the designer can layout a design. (See Figure 19-1.) These systems provide the designer with instant feedback as to the engineering feasibility of each layout. When the designer is satisfied that he has a completed design, he can instantly store the design and all necessary data in computer files. These files are accessible simultaneously to anyone who needs it for documation and manufacturing of P.C. boards.

Equipment shown in Figures 19-4 thru 19-10 courtesy of APPLICON, INC.

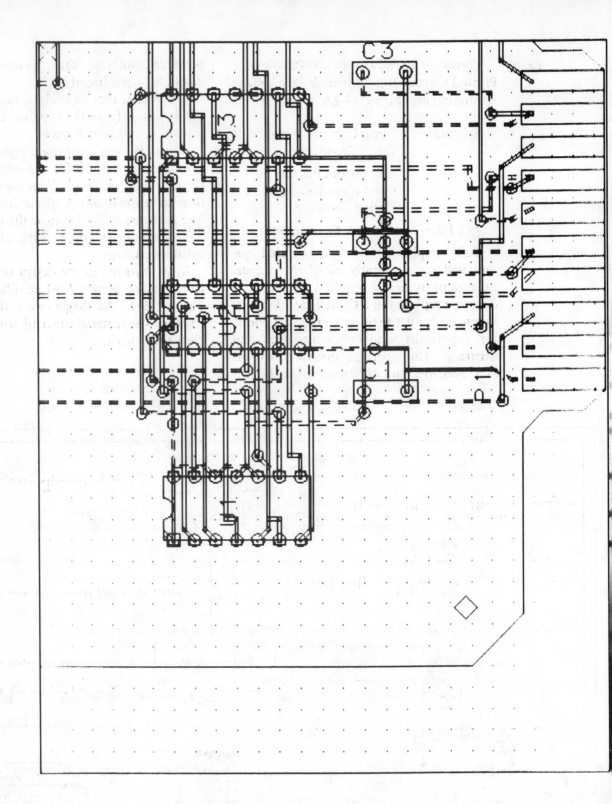

FIGURE 19-1

CADD (COMPUTER-AIDED DESIGN AND DRAFTING)

CADD Computer Aided Design and Drafting for Printed Circuits Boards includes the following functions: (Figure 19-2A, B and C)

Schematic   Logic Gate Pin Assignment
                     Logic Grouping

Layout   Component Placement
                     Conductor Routing
                     Design Checking

1x1 Film   Artwork Generation

Fully implemented, an automated system will replace nearly all of the manual operations typically required. However, it does not replace the P.C. Designer. Systems minimize layout time by combining operator experience with computer speed and accuracy. The operator oversees progress and overrides the computer when his experience indicates that another routing would be more desirable.

The practicality of using a computer depends primarily on the type and number of printed circuit boards made.

Normally, the computer software programs are developed for a particular computer system, to work from input (in the form of a signal list) or special instructions. The software is the brain of the computer. The computer can do nothing without the software package.

Steps involved in the design of a printed circuit board using CADD are illustrated in Figure 19-3. All begin with the design engineer's schematic and end with a complete final check.

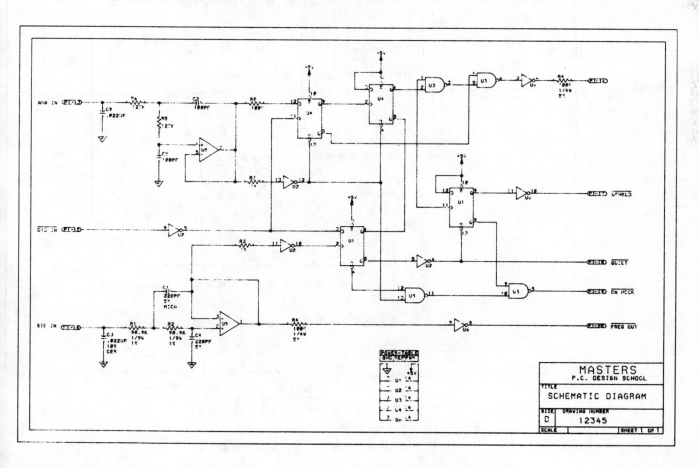

FIGURE 19-2A
Pen Plotted
Schematic

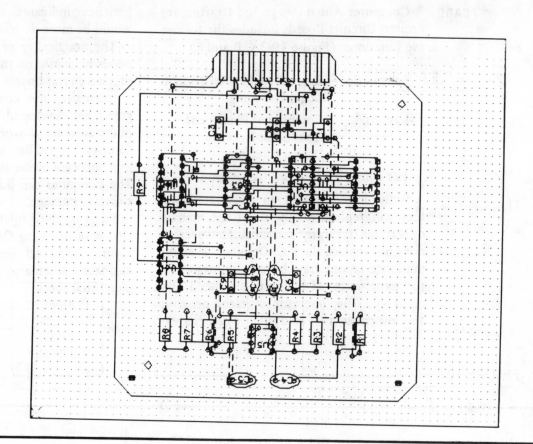

FIGURE 19-2B

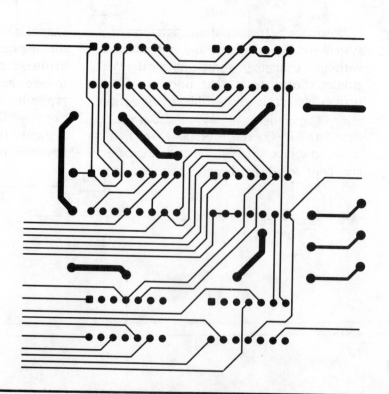

FIGURE 19-2C

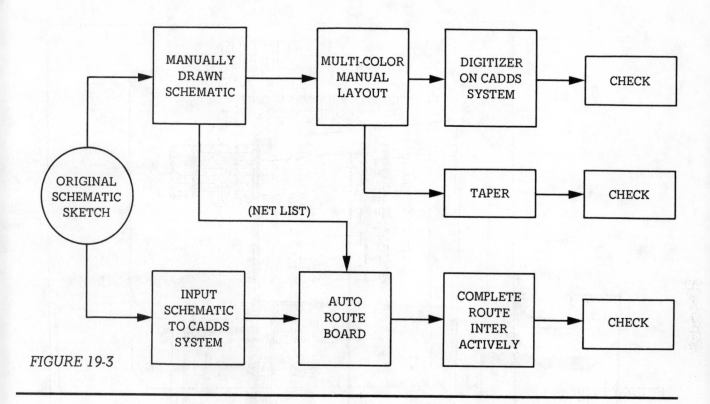

FIGURE 19-3

Hardware refers to those portions of a system which are fixed and not alterable without changing some physical electronics. Software is that portion of the system which is changeable by programming. The equipment in Figure 19-4 which shows a CPU (Central Processing Unit), disc drive, graphics display, function-keyboard are considered hardware. Only the pro-

gramming is considered software. Usually the software programmer is given a fixed hardware system and he proceeds to customize the software. The hardware would typically have certain limiting specifications. Additional CADD hardware is shown in Figure 19-5—Digitizing Table, Keyboard/ Printer Terminal, Line Plotter.

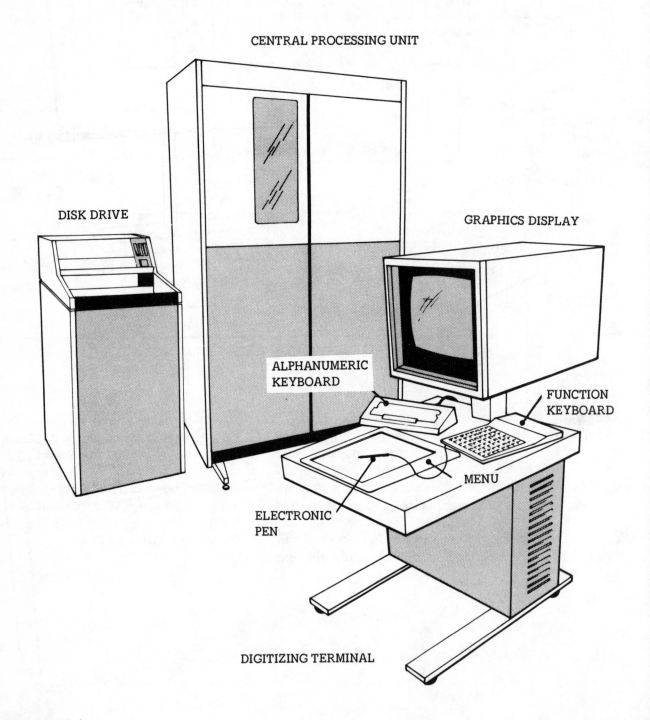

CENTRAL PROCESSING UNIT

DISK DRIVE

GRAPHICS DISPLAY

ALPHANUMERIC
KEYBOARD

FUNCTION
KEYBOARD

MENU

ELECTRONIC
PEN

DIGITIZING TERMINAL

FIGURE 19-4

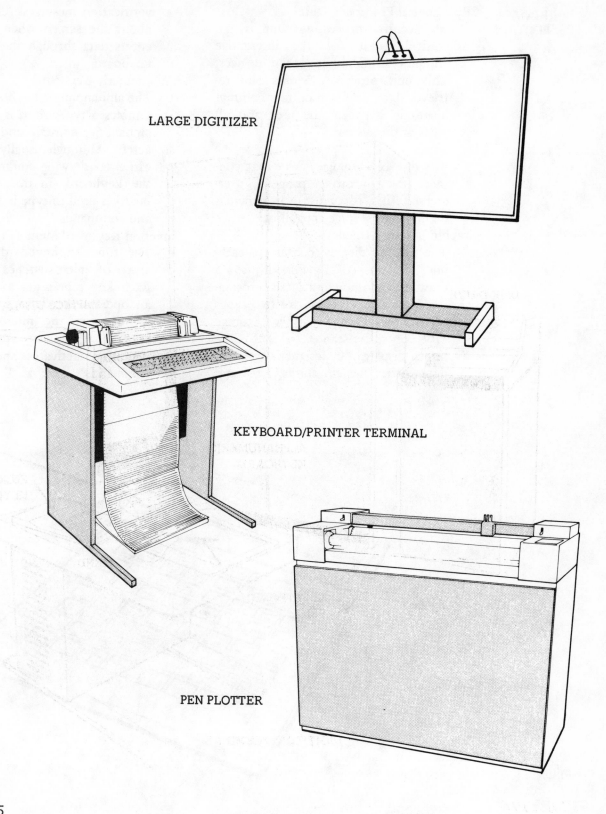

LARGE DIGITIZER

KEYBOARD/PRINTER TERMINAL

PEN PLOTTER

FIGURE 19-5

CPU Central Processing Unit

The central processing unit is the control point for the interactive graphics terminals and their devices. This unit processes, stores and retrieves data, and it monitors communications between the user and all parts of the system.

Disc Drive

The disc is a storage device that contains the operating program user-created files (drawings). It provides on-line mass storage capability.

Graphic Terminal Display

The graphic display contains a cathode ray tube (CRT), that displays a drawing and drawing-related information on a screen. There are two types of CRTs used for the display. Storage tube, or raster-scan tube. The CRT screen displays the designer's artwork, data entries, system status, error and verification messages. A cursor move about the screen when the operato enters data through the tablet or the keyboard.

Alphanumeric Keyboard

The alphanumeric keyboard, which re sembles a typewriter, is used to ente alphabetic, numeric and special char acters. Although usually the operato can enter drawing commands throug the keyboard, in this case the key board is used to type in text, number and commands.

Function Keyboard Menu (Figure 19-6)

The function keyboard Menu is a array of microswitches in a matrix Each key represents a function tha an operator predefines on the menu layout over the microswitches. By pressing a microswitch key, the operation activates specific system commands.

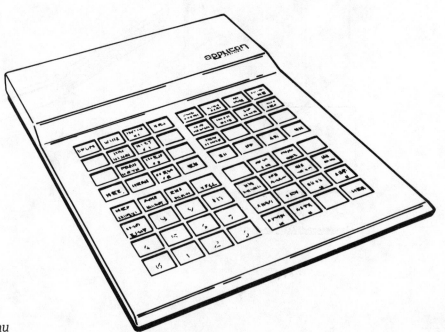

FIGURE 19-6
Function keyboard with menu

Digitizing Table and Control (Figure 19-7)

The table could take two types of pens—one for standard non-marking to draw symbols, the other a ballpoint for making permanent records of component outlines, symbols or graphic data. The digitizing control is a hand-held device which is used with the table. This control allows the operator to digitize the drawing or P.C. layout by depressing push-button switches which will enter data into the video display.

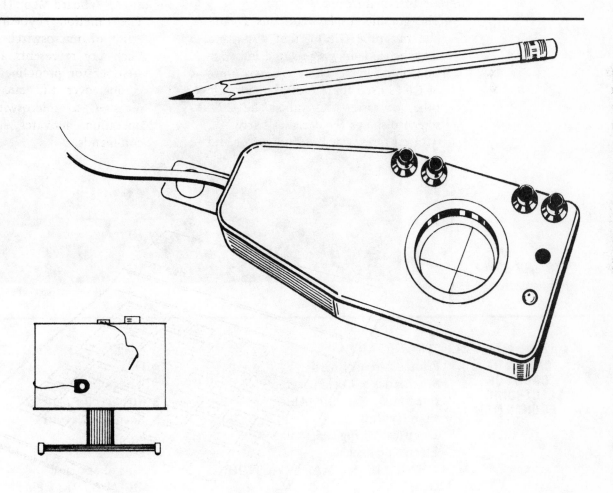

FIGURE 19-7
Digitizing control

Photoplotter

Once the P.C. Board is constructed and edited as desired, the photographic master film is produced. This photoplotting process consists of moving a controlled beam of light over a piece of film to produce connects, or traces; flashing the light while stationary will produce pads. Various types of plotters and photoheads are available. Various trace widths and pad sizes are produced by apertures in the photohead. The desired aperture is indicated during digitizing, and automatically selected during photoplotting.

PHOTOHEAD

FIGURE 19-8
Photoplotter

Photo courtesy Gerber Scientific

MAJOR APPLI-CATION OF CADD FOR ELECTRONIC EQUIPMENT

ELECTRONIC AREA:
  Printed Circuit Boards
  Schematics & Logic Diagrams
  Integrated Circuits—Thick & Thin
  Film Hybrids
  Electrical Wiring Diagrams
  Electronic Packaging.
MECHANICAL & MANUFACTURING AREA:
  Drafting and Documentation
  Sheet Metal Layout
  Tool Design
  Geometric Analysis
  3-Dimensional Design
  N.C. Program Generation

ARCHITECTURE AND ENGINEERING AREA:
  Design Drafting
  Structural Analysis
  Facilities Layout
  Piping
  Diagramming
  Bills of Material
PROGRAM DEVELOPMENT:
  In many cases, the necessary software is not included as part of the hardware, but must be purchased from a "Software House."

P.C. BOARD ON-LINE DESIGN VS OFF-LINE DESIGN

Hardware connections determine whether an on-line or off-line condition exists. A piece of equipment such as a photoplotter (Figure 19-8) is considered on-line or off-line, with respect to the computer (CPU). Direct hardware connections between the two enable the computer to pass information directly to the photoplotter. Such equipment is often connected to a temporary storage device. Data is generated by the computer, stored on magnetic tape and then transferred physically to the off-line tape drive for the photoplotter. On-line and off-line computer processing refers respectively to real-time and batch-processing.

BATCH PROCESSING—Operations such as a request to execute a particular program are often referred to as batch operations.

This type processing is generally adequate for many portions of the CADD process. For a P.C. Board on-line design break-down see Figure 19-9.

MAJOR APPLICATION OF CADD SOFTWARE

Design Check—Checks minimum width and spacings

Diagramming—Automatic Aids for Creating Diagrams

Geometric Construction—Graphic aids for constructing arcs and lines.

Auto Continuity—Checks a schematic of logic diagram to a layout.

Bill of Materials—List Components from a drawing.

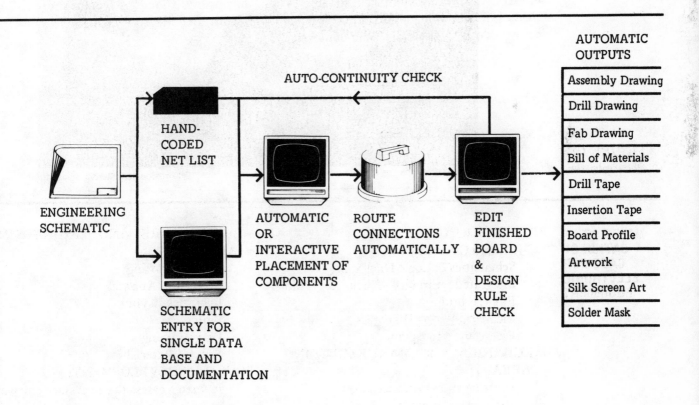

FIGURE 19-9
P.C. board on-line design

Component-placement-and density is perhaps the single most difficult parameter to generalize upon for designs supported by automated design techniques. Component density is a dependent parameter, affected in some degree by all of the process of CADD Design. Board size and proportions can be expected to directly affect conductor density. The concentration of conductors in certain portions of the board affect component density. The concentration of conductors or traces in the center of the board usually increases as the board size increases. Regarding board proportions, a 6x6 inch board will support a higher component density than a 3x9 inch board due to an improved balance of conductor or trace channels for side to side communication.

Grid Size has a direct bearing on the gross number of routing paths available. A smaller grid size will support a higher component density. Typically, the minimum practical grid size is determined by the desired conductor/trace width and the desired air gap or clearance between conductors. For example, a 20-mil conductor and a 30-mil airgap will result in a 50-mil grid size. Also a 12-mil conductor and a 13-mil air gap would require a 25-mil grid. The use of CADD automated design will often permit the use of smaller grid size than manual layout techniques would. This is a result of the high precision capabilities of CADD equipment, providing improved response to manufacturing tolerance requirements through a higher precision artmaster or film master. With the P.C. Board outline displayed on the CRT graphics display screen, the designer can call up parts from his data base and locate them with a light pen, thumb wheels, or "Joystick." The user selects basic construction, edit, storage, input, or retrieval functions from a keyboard or menu: (See Figure 19-10).

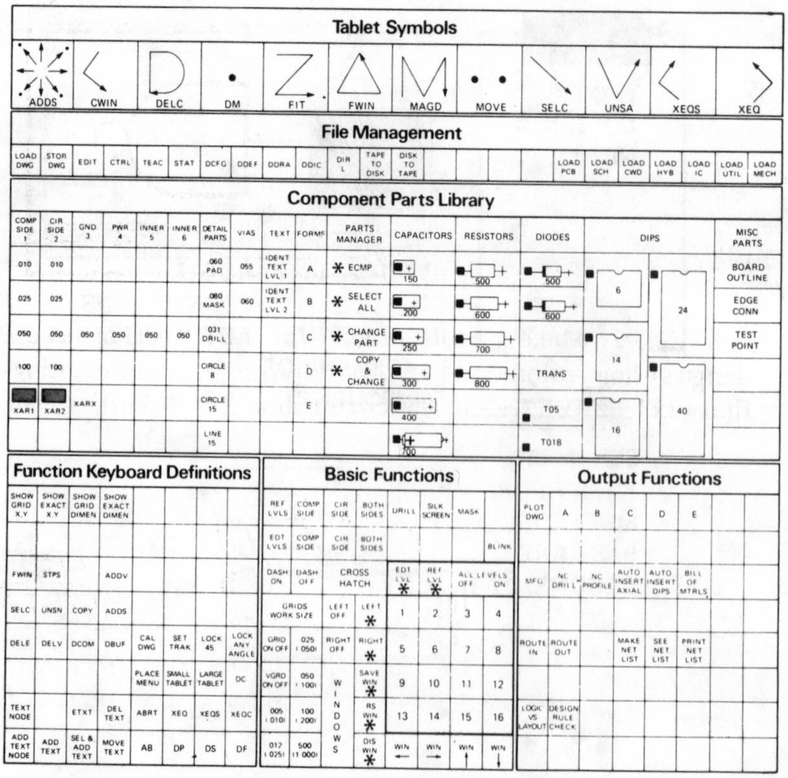

FIGURE 19-10
Example menu

SINGLE DATA BASE OUTPUTS (NC)
- N.C. Drill
- N.C. Profile
- N.C. Wirewrap
- N.C. Dip Insertion
- N.C. Axial Insertion

With a little extra effort in the preparation of layout data for artwork generation and some software development, numeric control data for various manufacturing equipment can be derived automatically. This has proved quite successful for drilling, wirewrap, component insertion. Major added advantage is the established data consistency which can be achieved in this manner.

DIGITIZING P.C. BOARDS

Digitizing is excellent for realizing productivity increases in a short period of time. The function of digitizing is to manually extract x-y coordinate information from the design layout through the use of a "DIGITIZER." During the process of digitizing, design layout errors may be discovered by the operator. The operator should be able to resolve these problems prior to the completion of digitizing. After digitizing the data, a pen plot is presented for checking. The final output (Film) is then completed by creating a photoplot. Depending on the complexity of the design, the digitizing steps can run from one hour to sixty hours or more. The following steps illustrate a typical digitizing process.

DIGITIZING PROCESS

STEP #1: Design the printed circuit in accordance with company or mil-standard.

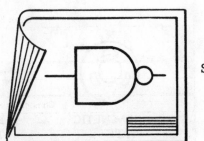

SCHEMATIC

STEP #2: Make a manual layout of the design using colored pencils on drafting film (2x1 or 4x1 scale). The circuit should be made using multicolor lines, each color representing a level/layer of the board on drafting film.

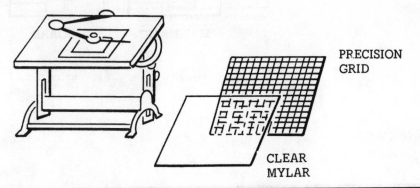

PRECISION GRID

CLEAR MYLAR

STEP #3: Digitize layout. Data will be stored on magnetic tape.

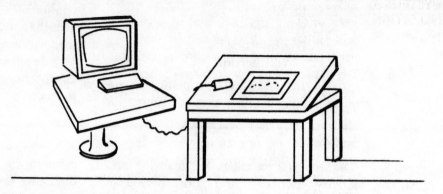

*After digitizing, the data is presented for checking; one form is a pen plot of the data.

STEP #4: Computer output using paper tape and magnetic tape will produce:

Assembly Drawing	Insert Tapes
Drill Drawing	Board Profile
Fab Drawing	Artwork
Bill of Materials	Silkscreen Art
Drill Tape	Solder Mask

MAGNETIC
TAPE

PAPER
TAPE

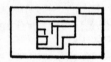

SCHEMATIC

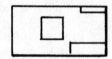

BOARD

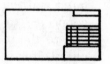

ASSEMBLY

TYPES OF SYSTEMS & UTILIZATION

When considering a CAD (computer aided design) system, you will find that it is very unlikely that any one type of system will satisfy all of your requirements. It is very rare to find a system that can satisfy cost and design requirements in one package. Many decisions concerning the ultimate and immediate extent of the system, the rate of system development, expansion, and mode of operation will have to be made. In all cases, the greater the automation of the system and its utility to the user, the greater the demands. Many large users have successfully employed multiple-interconnected computers with many access terminals to provide maximum utility for design. How the CAD system will be configured is dependent upon each company's need and growth requirements. Also, the effectiveness and efficiency of the CADD system can be greatly affected by the operating environment.

POTENTIAL CADD BENEFITS

Shorten production lead times

In analysis, make the recognition of component interactions easier

Help avoid subcontracting to meet schedules

Make the management of design manpower more effective

Enrich the engineering organization with technical cross-fertilization

Provide better cost control

Lead to easier customer modification

Provide the potential for using more existing parts and tooling

Help ensure designs appropriate to existing manufacturing techniques

TIME SHARING

The complexity of integrated circuits is forcing P.C. Designers to turn to computer aided design. For a small company, however, the cost of a typical CAD system ($250,000 for example) could be prohibitive. However, there are less expensive alternatives: companies now offer CAD service over telephone lines at low cost. You may buy or rent a digitizing terminal and use a remote computer with high speed peripherals. The typical configuration would be a storage display, keyboard, computer and floppy disk. The work station is connected to the service bureau computer and to the required high performance peripherals for plotting and high speed printout.

Time sharing systems essentially reflect the ability of a computer to be switched from one job to another quickly. It always keeps track of where it has been and where it is going; is able to come back to any interrupted task and continue execution from the last executed operation. Consequently, time on the computer can be optimized by providing the proper instructions that will tell the computer to switch to another task when delays are encountered. Delays occur frequently, especially where large volumes of input and/or output are encountered. For applications where the input is provided through devices such as light pens or typewriter keyboards, these delays can be extensive.

Many terminals can be serviced by the local computer in a manner that appears to each user that he has complete and immediate access to the computer. When time sharing is coupled with a communications facility that allows individual terminals to be placed at remote sites, the result can be a high degree of flexibility for design communication.

SERVICE BUREAUS OFFERING P.C. DESIGN

Service Bureaus can give a good combination of computer efficiency and manual control to guarantee that your finished artwork will be correct and delivered on time. Most service bureau P.C. Design consultants can help you select what design rules you need. They should explain the design trade-offs with respect to speed and manufacturability. Data is entered from your schematics; the board is designed and checked by the service bureaus to your input schematic. This work is typically guaranteed; if the finished artwork does not conform completely to your specified input, they correct it. Typically all data on each P.C. Board is maintained in their library so changes can be made quickly and easily.

SERVICE BUREAU TYPICAL REQUIRED INPUT

For Quotation:
1. Schematic or Wire List
2. Board Outline
3. Design Rules
4. Required Turnaround

For Job Completion:
1. Schematic or Net List
2. Board outline with dimension of all features.
3. Spec sheet for all non standard components
4. Spec sheet of all connectors used on P.C. Board.
5. List of deliverables such as:
   A. Drill Tape
   B. Ground Planes
   C. Film Negatives
   D. Fabrication Drawing
   E. Assembly Drawing
   F. Schematic
   G. Silkscreen Film & Drawing

The following is a brief look at most of the CADD systems on the market today that will support printed circuit design. The home office address is listed along with the model of the processor used, and any other applications that are supported besides P.C. design.

Photo Courtesy of Applicon, Inc.

Photo Courtesy of CALMA

Photo Courtesy of Redac Interactive Graphics, Inc.

Photo Courtesy of Computervision Corporation

APPLICON, INC.
32 Second Avenue
Burlington, MA 01803

Digital Equipment Corp.
PDP 11/34

Mechanical, Electrical, Architectural, Civil Engineering, Facilities Layout

AUTO-TROL TECHNOLOGY CORPORATION
5650 Pecos St.
Denver, CO 80221

Sperry-Univac V77-600

Mechanical, Electrical, Architectural, Civil Engineering, Facilities Layout

CALCOMP (California Computer Products)
3320 E. La Palma Ave., P.O. Box 4407
Anaheim, CA 92803

Calcomp # CC 16-40
Minicomputer

Mechanical, Electrical, Architectural, Civil Engineering, Facilities Layout

CALMA
527 Lakeside Drive
Sunnyvale, CA 94086

Data General 2-230

Mechanical, Electrical, Architectural, Civil Engineering, Facilities Layout

COMPUTERVISION CORPORATION
201 Burlington Road, (Route 62)
Bedford, MA 01730

Computer Vision Corp.
CGP-100 & CGP-200

Mechanical, Electrical, Architectural, Civil Engineering, Facilities Layout

DATA TECHNOLOGY, INC.
4 Gill Street
Woburn, MA 01801

DEC LSI 11/00 Series

Mechanical, Electrical, Architectural, Civil Engineering,
Facilities Layout

GERBER SCIENTIFIC INSTRUMENT CO.
83 Gerber Road
South Windsor, CT 06074

Hewlett Packard
Models 2112, 2113 & 2117

Mechanical, Electrical Architectural, Civil Engineering,
Facilities Layout

IDI (Information Displays, Inc.)
150 Clearbrook Road
Elmsford, NY 10523

Sperry-Univac V77-600

Mechanical, Electrical, Architectural, Civil Engineering,
Facilities Layout

INTEGRAPH CORP.
One Madison Industrial Park
Huntsville, AL 35807

Digital Equipment Corp.
Models 11-34, 11/70, 11/780

Mechanical, Electrical, Architectural, Civil Engineering,
Facilities Layout

NICOLET ZETA CORPORATION
2300 Stanwell Drive
Concord, CA 94520

Nicolet Zeta
Model 6300

Mechanical, Electrical, Architectural, Civil Engineering,
Facilities Layout

REDAC INTERACTIVE GRAPHICS INC.
One Redac Way
Littleton, MA 01460

Digital Equipment Corp.
PDP 11/34

SUMMAGRAPHICS CORPORATION
35 Brentwood Ave.
Fairfield, CT 06430

Data General
Micronova

Mechanical, Electrical, Architectural, Civil Engineering,
Facilities Layout

Chapter

20

HELPFUL

HINTS

This chapter offers a collection of helpful hints and suggestions regarding the design of printed circuit boards.

Use it as a checklist. It covers the most salient design considerations with regard to component density, mounting, grounding, bypassing and the disposition of unused logic gates. This concluding chapter highlights a wide variety of design hazards, manufacturing processes and safety responsibilities of the PC board designer, and it consolidates thermal, frequency and shielding considerations for PC boards.

There are specific recommendations included with regard to the three basic logic technologies: CMOS, TTL and ECL.

These fabrication techniques represent the three basic speed/power tradeoffs available to electronic design engineers: CMOS stands for "complementary metal oxide semiconductor" and it is the lowest power-consuming, slowest design option. TTL (or T^2L) stands for "transistor transistor logic" and it is the mid-range speed/power trade-off. ECL stands for "emitter-coupled logic" and it is the fastest, hottest logic technology. There are other technologies to be sure, but these three represent 99^{44} /100% of all commercial applications.

DENSITY
IC per sq. inch

LIMITATIONS

- CIRCUIT DESIGN
- MULTILAYER BOARD OR TWO SIDED
- PACKAGING RESTRICTIONS
- P.C. BOARD DESIGNER
  - EXPERIENCE
  - CAPABILITIES
  - WORK PRESSURES
- CONDUCTOR WIDTH/SPACING

  IF GREATER THAN .012
  DENSITY GOES DOWN
  YIELD GOES UP

  IF LESS THAN .012
  DENSITY GOES UP
  YIELD GOES DOWN

PROCESSING

- CLEANING
- SOLDER MASK
- CONFORMAL COATING
- COMPONENT REMOVAL
- LEGENDS (COMPONENT IDENT.)

CLEANING—CLEAN FOR ORGANICS
SOLDER MASK—CURED EPOXY
CONFORMAL COATING
  POLYURETHANES, EPOXY, SILICONES, ACRYLICS

COMPONENT REMOVAL
  HEAT & PULL—THEN PRAY, IMPULSE SUCKER, VACUUM DESOLDERING

LEGEND—COMPONENT IDENTIFICATION SYMBOLS, REF. DESIGNATIONS

METALLIC GROWTHS
DENDRITIC GROWTHS

CONDUCTORS

CONTRIBUTORS:

ENVIRONMENTAL—HUMIDITY GREATER THAN 70% DESIGN—

- CONTINUOUSLY APPLIED VOLTAGES GREATER THAN 3 VDC
- CLOSE CONDUCTOR SPACING
- METAL SELECTIONS
- SHARP CORNERS
- CONTAMINATION

METALLIC GROWTHS
CONDUCTIVE CONTAMINATION

CONTRIBUTORS
RESIDUES FROM—

- PLATING
- ETCHING
- FUSING
- SOLDERING (FLUX)
- FINGERPRINTS
- ATMOSPHERE
- UNCURED RESINS

ALL COMPOUNDED
BY INEFFECTIVE
CLEANING
PROCEDURE—

DENDRITIC GROWTHS

RESULT IN LEAKAGE CURRENTS UP TO 40 MA @ 5 VDC WITH 0.5 MM CONDUCTOR SPACING (.020)s,

WHISKER GROWTHS

HAIRLIKE SINGLE
XTAL WHISKER

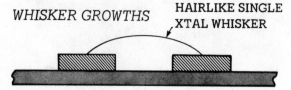

THERMAL

A PRINTED BOARD SHOULD NOT BE USE
AS A "HEAT SINK" COMPONENTS DISS
PATING MORE THAN 1 WATT OR HAVIN
A SURFACE TEMPERATURE GREATE
THAN 120°C SHOULD NOT BE MOUNTE
IN CONTACT WITH THE BASE MATERIA
UNLESS HEAT DISSIPATION DEVICE
SUCH AS HEAT SINKS OR THERMA
GROUND PLANES ARE USED.

IDENTIFY HAZARDS

THINK THROUGH WHAT MIGHT POSSIBL
GO WRONG, AND WHAT THE CONS
QUENCES WOULD BE IF IT DID.

DESIGN FOR SAFETY

CONSIDER SAFETY TO BE AN INTEGRA
PART OF THE DESIGN PROCESS. PROTE
TIVE DEVICES, WARNING SIGNS, AN
PROCEDURES ARE SUPPLEMENTS T
GOOD DESIGN BUT CAN NEVER FULL
COMPENSATE FOR ITS ABSENCE. IT I
THE RESPONSIBILITY OF THE P.C. BOAR
DESIGNER TO:

1. IDENTIFY POTENTIAL HAZARDS
2. ISOLATE THOSE HAZARDS
3. PROVIDE PROTECTIVE BARRIER
   WHEN REQUIRED

LIVE PART

ANSI C39.5-1974 DEFINES A "LIVE PART
AS AN ELECTRICALLY CONDUCTIN
PART CARRYING MORE THAN 30VAC O
42.4 VDC OR PEAK

NOT LIVE

NOT LIVE IF CURRENT FROM THE PAR
TO GROUND THROUGH 1500 OHMS CAI
NOT EXCEED 0.5 MILLIAMPERES, EVE
THOUGH THE PART CARRIES VOLTAG
EQUAL TO OR GREATER THAN THA
SPECIFIED FOR A "LIVE PART"

HELPFUL HINTS

CONDUCTIVE PATTERNS

CAREFUL CONSIDERATIONS MUST BE GIVEN TO THE DESIGN OF THE CONDUCTIVE PATTERNS. A GOOD PRINTED CIRCUIT BOARD DESIGN INCLUDES MANUFACTURABILITY, REPAIR/MAINTENANCE CONSIDERATIONS, AND ALL FUNCTIONAL REQUIREMENTS.

PRIMARY CONCERNS

IN THE DESIGN PHASE, THE PRIMARY CONCERNS ARE THE DESIGN WIDTH OR CONDUCTORS, CONDUCTOR SPACING, COMPONENT HOLES, THROUGH CONNECTIONS, & COMPONENT MOUNTING.

COMMENT

A GOOD PRINTED CIRCUIT BOARD DESIGN ENTAILS THE JUDICIOUS USE OF THE PREFERRED "MINIMUM" & MINIMUM BY EXCEPTION. ON CONDUCTOR WIDTHS, CONDUCTOR SPACING, & LAND DIAMETERS. MINIMUM DESIGN SIGNIFICANTLY INCREASES THE COST OF MANUFACTURING PRINTED CIRCUIT BOARDS AS WELL AS REDUCING THE FUNCTIONAL RELIABILITY.

MINIMUM'S:

WHEN THE PRINTED CIRCUIT BOARD DESIGN REQUIRES THE PREFERRED "MINIMUM" AND MORE IMPORTANTLY THE "MIN. BY EXCEPTION" CONDUCTOR SPACING, SERIOUS CONSIDERATION SHOULD BE GIVEN TO REQUIRING THE PRINTED BOARD ASSEMBLY TO BE CONFORMALLY COATED OR A SUITABLE SOLDER MASK APPLIED.

Note: WITH PRESENT STATE-OF-THE-ART CAPABILITIES IT IS POSSIBLE TO USE SMALLER LAND DIAMETERS, ANNULAR RINGS AND HOLE DIAMETERS. HOWEVER, THESE SMALLER GEOMETRIES SIGNIFICANTLY INCREASE MFG. COST AND REDUCE MANUFACTURABILITY.

- WHEN WIDE CONDUCTORS MUST BE TRANSFERRED FROM ONE CONDUCTIVE LAYER TO ANOTHER, USE MORE THAN ONE PLATED-THROUGH-HOLE AS SHOWN.

USE OF MORE THAN ONE PLATE-THROUGH-HOLE TO CONNECT WIDE CONNECTORS.

- CONDUCTOR SPACING
THE CONDUCTOR SPACING SHALL BE ESTABLISHED FOR EACH PRINTED CIRCUIT BOARD DESIGN'
  - RECOMMENDED CONDUCTOR SPACING
    - *ROUTINE DESIGNS:* .025 MIN. VOLTAGE LIMITATIONS 400 VDC OR AC PEAK. (SEE SAFETY REQUIREMENTS HIGH VOLTAGE)
    - *HIGH DENSITY:* THE "PREFERRED MINIMUM" IS .015, AND THE VOLTAGE LIMITATION IS 50 VDC OR PEAK.
    - THE "MINIMUM BY EXCEPTION" is .015 OR AC PEAK

LOGIC SPEED

NAME	NUMBER	SPEED	POWER PER GATE
LOW POWER	74LOO	3 MHZ	1 MW
STANDARD	7400	25 MHZ	10 MW
LOW POWER SCHOTTKY	74LSOO	25 MHZ	2 MW
STANDARD SCHOTTKY	74SOO	125 MHZ	19 MW
STANDARD HIGH PWR	74HOO	50 MHZ	22 MW
ECL	–	100 MHZ	20 MW
C–MOS	–	3 MHZ	VERY LOW .01 MW

- CRITICAL LEADS CAN BE IMPROVED BY MAKING THE TRACE WIDER OR SHORTER.
- IF STANDARD IC'S & HIGH SPEED ON THE SAME BOARD – BREAK UP GROUND . . .

POWER DISTRIBUTION RECOMMENDATIONS

T^2L

- LOW FREQUENCY SERIES—CONNECTED COMMON GROUND SYSTEM
- USE A 1uf to 10uf TANTALUM CAPACITOR NEAR THE INPUT
- USE A .01uf TO .1uf CAPACITOR FOR EVERY 1-5 IC's

C-MOS

- LOW FREQUENCY SERIES—CONNECTED COMMON GROUND SYSTEM
- MIXING C-MOS & T^2L—QUESTION TO ASK THE ENGINEER? SHOULD I SEPARATE THE C-MOS & T^2L GROUND ON THE P.C. BOARD

C-MOS

- USE A SMALL VALUE SERIES RESISTOR, ZENER DIODE AND A CAPACITOR TO GROUND ON EACH P.C. BOARD FOR EVERY 1 - 100 IC'sS.

ECL

- HIGH FREQUENCY—USE A GROUND PLANE WITH MULTIPOINT GROUNDING.
- MULTILAYER BOARD WOULD BE BEST.
- ECL—GROUPING & PLOTING IS VERY IMPORTANT
    - *CRITICAL LEADS CAN BE IMPROVED BY MAKING THE TRACE WIDER.
- IF ECL IS MIXED YOU SHOULD HAVE A COMMON GROUND PLANE & A VOLTAGE PLANE.

ECL

- USE A .01uf TO .1uf CAPACITOR FOR EVERY 1-5 IC'S.
- USE A 1uf TO 10uf TANTALUM NEAR THE INPUT CONNECTOR PIN.

SCHOTTKY

- HIGH FREQUENCY—USE A GROUN PLANE WITH MULTIPOINT CONTAC GROUNDING AT THE INPUT CONNEC TOR.
- MULTILAYER BOARD WOULD BE BES
- IF SCHOTTKY & STANDARD TTL AR MIXED YOU SHOULD HAVE SEPARAT WIDE GROUNDS.
- USE A .01uf TO .1uf CAP. FOR EVER 1-5 IC'S.
- USE A 1uf TO 10 uf TANTALUM NEA THE INPUT CONNECTOR.

UNUSED INPUT GATES

T^2L

TIE TO USED INPUT OF THE SAME GAT (WATCH LOADING). TIE TO VC THROUGH A 1 K Ω RESISTOR FOR EVER 1 TO 25 UNUSED INPUTS.

LS

TIE TO VCC THROUGH A 1K Ω RESISTO FOR 1-25 UNUSED INPUTS. DO NOT TIE T A USED INPUT OF THE SAME GATE LOOSES SPEED.

C-MOS

TIE TO VDD, GND OR USED INPUT CHEC LOADING. ALL INPUTS "MUST" BE CON NECTED TO AN APPROPRIATE SUPPL VOLTAGE, OR ANOTHER USED INPUT.

POWER DISTRIBUTION

POWER DISTRIBUTION-BYPASS CAPACITOR LOCATIONS.

THE LOCATIONS OF THE BYPASS CAPACITOR FOR EACH INTERGRATED CIRCUIT CAN BE CRITICAL FOR PROPER FUNCTIONAL OPERATION OF THE DESIGN.

ACCEPTABLE

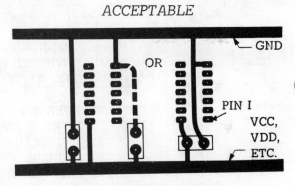

GND

OR

PIN I

VCC,
VDD,
ETC.

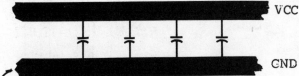

VCC

GND

POWER DISTRIBUTION BUS SYSTEM COULD BE A SHOCK EXCITED VHF RESONATOR.

- ADDITIONAL DECOUPLING CAP'S WILL TEND TO SHIFT THE RESONANT FREQUENCY.

- POSSIBLE SOLUTION: USE AN ADDITIONAL LOSSY CHOKE AND CAPACITOR FOR EACH (OR PAIR) OF IC'S.

UNACCEPTABLE

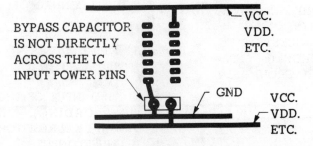

BYPASS CAPACITOR IS NOT DIRECTLY ACROSS THE IC INPUT POWER PINS

VCC.
VDD.
ETC.

GND

VCC.
VDD.
ETC.

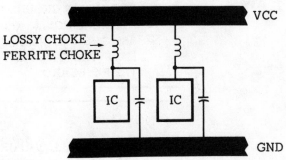

LOSSY CHOKE
FERRITE CHOKE

VCC

IC   IC

GND

NON-PREFERRED BUS STRUCTURE

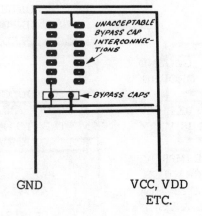

UNACCEPTABLE BYPASS CAP INTERCONNECTIONS

BYPASS CAPS

GND   VCC, VDD
                       ETC.

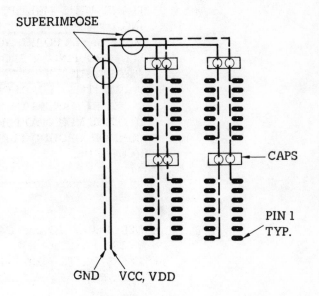

SUPERIMPOSE

CAPS

PIN 1
TYP.

GND   VCC, VDD

ACCEPTABLE BUS STRUCTURE

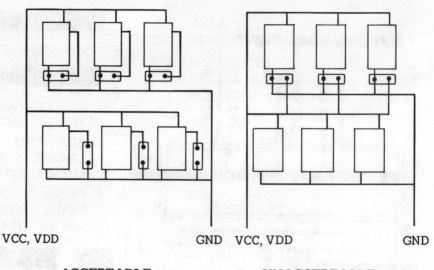

VCC, VDD   GND   VCC, VDD   GND

ACCEPTABLE   *UNACCEPTABLE*
BUS STRUCTURE   *BYPASS CAPACITOR*
                                  INTERCONNECTION

POWER DISTRIBUTION CONDUCTIVE
PATTERNS WITH CAPACITIVE
DECOUPLING

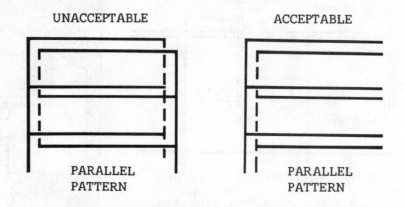

UNACCEPTABLE   ACCEPTABLE

PARALLEL   PARALLEL
PATTERN   PATTERN

PREFERRED BUS STRUCTURE

ALL SUPERIMPOSED CONDUCTORS SHOULD BE 2 MM (.080) TO 4 MM (.160). PARALLEL CONDUCTORS SHOULD BE 1.5 MM (.060) TO 3 MM (.120)

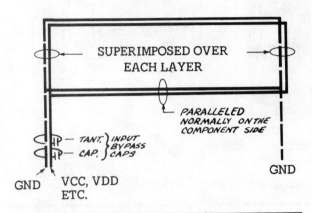

ACCEPTABLE BUS STRUCTURE

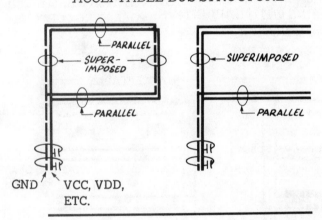

GROUNDING NOTES:

1. GROUNDING IS ONE OF THE PRIMARY WAYS TO MINIMIZE UNWANTED NOISE AND PICKUP OF UNDESIRED SIGNALS.
2. A GOOD GROUND SYSTEM MUST BE DESIGNED.
3. THERE ARE TWO BASIC OBJECTIVES:

   A. MINIMIZE THE NOISE VOLTAGE GENERATED BY CURRENTS FROM TWO OR MORE CIRCUITS WHICH ARE FLOWING THROUGH A COMMON CONDUCTOR.
   B. AVOID CREATING GROUND LOOPS WHICH ARE SUSCEPTIBLE TO MAGNETIC FIELDS.

GROUND PLANE DESIGN CONSIDERATIONS

- SEPARATE GROUNDING FOR:
    SIGNAL (SENSITIVE)
    POWER (HIGH CURRENT)
    CHASSIS
- IDEALLY A COMPLETE LAYER
- SECTIONS OF A MODIFIED GROUND PLANE SHOULD NOT BE ISOLATED AND CONNECTED BY A "LONG" NARROW CONDUCTOR.

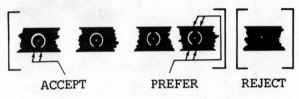

GROUND PLANE DESIGN CONSIDERATION – (MOSTLY FOR ANALOG DESIGN)

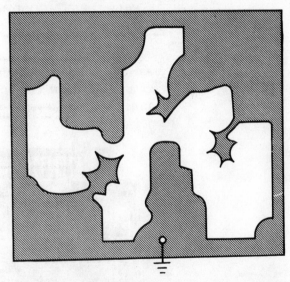

ELECTROSTATIC SHIELDING FOR LOW POWER SIGNALS—NOT SUITABLE FOR HIGH CURRENT OUTPUT W/OUT SUITABLE GROUND RETURN.

A ground plane should not be a long serpentine of conductive areas interconnected by long thin sections of conductors.

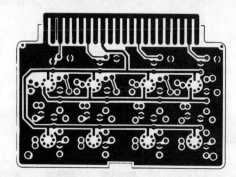

- Non-preferred ground (voltage) plane designs
  - GROUND (VOLTAGE) CONDUCTIVE AREAS SHOULD NOT BE INTERCONNECTED BY LONG THIN CONDUCTORS.
  - THE MANY SHARP POINTS AND CORNERS INCREASE THE POSSIBILITY OF SHORTS.
  - ALL GROUND (VOLTAGE) PLANES SHALL HAVE GOOD RF DECOUPLING.

A ground plane should consist of large areas of continuous conductive area, and should be brought out on the edge-board connectors to several equally spaced printed contacts.

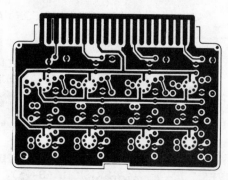

- Preferred ground (voltage) plane design.
  - LARGE AREAS OF CONTINUOUS CONDUCTIVE AREAS.
  - GROUND (VOLTAGE) PLANES BROUGHT OUT TO SEVERAL EQUALLY SPACED PRINTED CONTACTS AT THE EDGE BOARD CONNECTOR.

COMPONENT LEGEND LOCATIONS

The main objective of printing legends on P.C. board: To provide the information required to manufacture, assemble, functional test, maintain and identify the individual components and the P.C. board assembly.

The location and orientation of the component legend is not for the convenience of the P.C. designer.

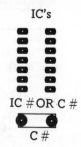

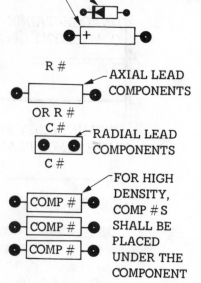

PREFERRED COMPONENT LEGEND

HELPFUL HINTS

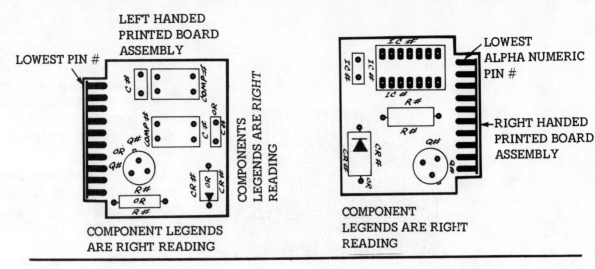

LEFT HANDED
PRINTED BOARD
ASSEMBLY

LOWEST PIN #

COMPONENTS
LEGENDS ARE RIGHT
READING

COMPONENT LEGENDS
ARE RIGHT READING

LOWEST
ALPHA NUMERIC
PIN #

RIGHT HANDED
PRINTED BOARD
ASSEMBLY

COMPONENT
LEGENDS ARE RIGHT
READING

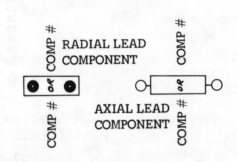

RADIAL LEAD
COMPONENT

AXIAL LEAD
COMPONENT

UNACCEPTABLE
COMPONENT LEGEND LOCATIONS

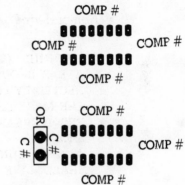

ACCEPTABLE
COMPONENT LEGEND LOCATIONS

TERMINOLOGY FOR DESCRIBING MODUAL
REFERENCE PLANES

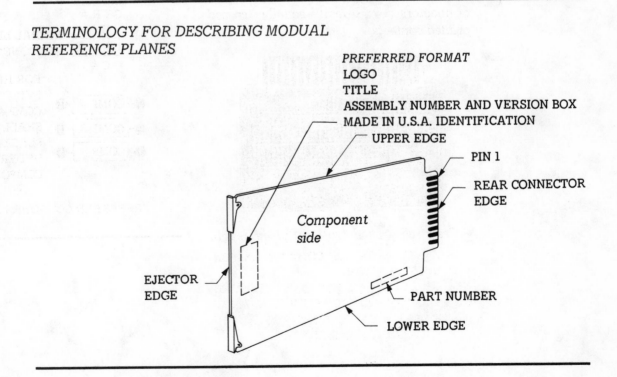

PREFERRED FORMAT
LOGO
TITLE
ASSEMBLY NUMBER AND VERSION BOX
MADE IN U.S.A. IDENTIFICATION
UPPER EDGE

PIN 1

REAR CONNECTOR
EDGE

Component
side

EJECTOR
EDGE

PART NUMBER

LOWER EDGE

HELPFUL HINTS

355

Appendix

DESIGN REFERENCE MATERIALS

TERMS APPLICABLE TO PRINTED CIRCUIT BOARDS

1. **AIR GAP:** The non-conductive air space between traces, pads, conductive matter, or any combination thereof.

2. **ANALOG CIRCUIT:** A circuit comprised mostly of discrete components (i.e., resistors, capacitors, transistors) which produces data represented by physical variables such as voltage, resistance, rotation, etc.

3. **ANNULAR RING:** The width of the conductor surrounding a hole through a Printed Circuit Pad.

4. **ARTWORK:** An accurately scaled configuration used to produce a Master Pattern. Generally prepared at an enlarged scale using various width tapes and special shapes to represent conductors.

5. **AWG:** American Wire Gage. A method of specifying wire diameter. The higher the number, the smaller the diameter.

6. **AXIAL LEADS:** Leads coming out the ends and along the axis of a resistor, capacitor, or other axial part, rather than out the side.

7. **BRIDGING:** A condition that generally happens during the wave soldering operation where excess solder builds up and shorts out adjacent conductors.

8. **BUS:** A heavy trace or conductive metal strip on the Printed Circuit Board used to distribute voltage, grounds, etc., to smaller branch traces.

9. **BYPASS CAPACITOR:** A capacitor used for providing a comparatively low impedance A-C path around a circuit element.

10. **CONFORMAL COAT:** A coating that is generally sprayed, dipped, or brushed on to provide the completed Printed Circuit Board protection from fungus, moisture and debris.

11. **CONNECTOR TONGUE:** A protrusion of the Printed Circuit Board edge that is manufactured to a configuration to mate with a receptacle that provides electrical and/or mechanical junction between the Printed Circuit Board and other circuitry.

12. **DIGITAL CIRCUIT:** A circuit comprised of mostly integrated circuits which operates like a switch (i.e., it is either "ON" or "OFF").

13. **DISCRETE COMPONENT:** A component which has been fabricated prior to its installation (i.e., resistors, capacitors, diodes and transistors).

14. **FEED-THRU:** A plated-thru hole in a Printed Circuit Board that is used to provide electrical connection between a trace on one side of the Printed Circuit Board to a trace on the other side. Since it is not used to mount component leads, it is generally a small hole and pad diameter.

15. **FLOW SOLDERING:** Also called wave soldering. A method of soldering Printed Circuit Boards by moving them over a flowing wave of molten solder in a solder bath.

16. **GLASS EPOXY:** A material used to fabricate Printed Circuit Boards. The base material (fiberglass) is impreginated with an epoxy filler which then must have copper laminated to its outer surface to form the material required to manufacture Printed Circuit Boards.

17. **GRID:** A two-dimensional network consisting of a set of equally spaced parallel lines superimposed upon another set of equally spaced parallel lines so that the lines of one set are perpendicular to the lines of the other.

18. **GROUND PLANE:** A condition where all unused areas (areas not consumed by traces or pads) of the Printed Circuit Board are left unetched and tied to the ground circuit throughout the board.

19. **MASTER PATTERN:** An accurately scaled pattern which is used to produce the Printed Circuit within the accuracy specified in the Master Drawing.

20. **MOTHER BOARD:** Also called Back Plane, or Matrix Board. A relatively large Printed Circuit Board on which modules, connectors, sub-assemblies or other Printed Circuit Boards are mounted and inter-connections made by means of traces on the board.

21. **PLATING:** A uniform coating of conductive material upon the base metal of the Printed Circuit Board

22. **RADIAL LEAD:** A lead extending out the side of a component, rather than from the end.

23. **REGISTRATION:** The alignment of a pad on one side of the Printed Circuit Board (or layers of a multi-layer board) to its mating pad on the opposite side.

24. **TOOLING HOLE:** Also called Fabrication Hole, Pilot Hole, or Manufacturing Hole.

DESIGN REFERENCE MATERIALS

RESISTOR AND CAPACITOR STANDARD COLOR CODE

STANDARD COLOR CODE FOR RESISTORS
AND CAPACITORS

The standard color code provides the necessary information required to properly identify color coded resistors and capacitors. Refer to the color code for numerical values and the number of zeros (or multiplier) assigned to the colors used. A fourth color band on resistors determines the tolerance rating. Absence of the fourth band indicates a 20% tolerance rating.
(REF MIL-STD-221)

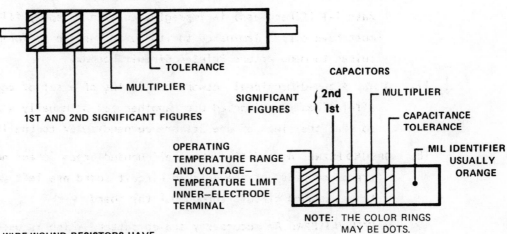

AXIAL LEAD RESISTOR

TOLERANCE
MULTIPLIER
1ST AND 2ND SIGNIFICANT FIGURES

CAPACITORS

SIGNIFICANT FIGURES { 2nd 1st }
MULTIPLIER
CAPACITANCE TOLERANCE
MIL IDENTIFIER USUALLY ORANGE
OPERATING TEMPERATURE RANGE AND VOLTAGE— TEMPERATURE LIMIT INNER—ELECTRODE TERMINAL

NOTE: THE COLOR RINGS MAY BE DOTS.

WIRE-WOUND RESISTORS HAVE
FIRST DIGIT BAND DOUBLE WIDTH

COLOR CODE

COLOR	FIRST FIGURE	SECOND FIGURE	MULTIPLIER	TOLERANCE %	LTR
BLACK	0	0	1		
BROWN	1	1	10		
RED	2	2	100	± 2%	G
ORANGE	3	3	1,000		
YELLOW	4	4	10,000		
GREEN	5	5	100,000		
BLUE	6	6	1,000,000		
PURPLE (VIOLET)	7	7			
GRAY	8	8			
WHITE	9	9			
SILVER			0.01	± 10%	K
GOLD			0.1	± 5%	J
				± 20%	M
				± 1%	F

OHM'S LAW EQUATIONS

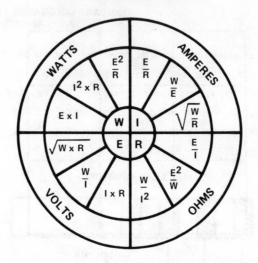

```
W = Power in Watts
E = Electro-Motive Force in Volts
I = Current in Amperes
R = Resistance in Ohms
```

```
Joule's Law, W = E x I, combined with
Ohm's Law gives the above 12 equations.
```

CONVERSION OF ELECTRICAL COMPONENT VALUES

CAPACITORS	
1,000,000 Microfarads (UF) = 1 Farad	
1,000,000 Picofarads (PF) = 1 Microfarad (UF)	
RESISTORS	
1,000 ohms = 1 k	
1,000,000 ohms = 1 Megohm (Meg)	
VOLTS & AMPERES	
1000 Millivolts = 1 Volt	1000 Milliamps = Amp
1000 Microvolts = 1 Millivolt	1000 Microamps = 1 Milliamp
INDUCTANCE	
1000 Millihenries = 1 Henry	
1000 Microhenries = 1 Millihenry	

WIRE CURRENT CAPABILITIES CHART

AWG	FUSING CURRENT (AMPS)	NORMAL LOAD (AMPS)
26	20	0.6
24	29	1.0
22	41	1.6
20	58	2.5
18	83	4.0
16	117	6.0
14	166	15.0
12	235	20.0

1. THE FUSING CURRENT is the current at which the wire will melt.

2. THE NORMAL LOAD (wire rating) is based on 400 circular mils per ampere.

3. Stranded wire specifications allow strands to run from 7 to 19 strands but manufacturers usually use 7 or 19.

How To Use This Chart: Locate the conductor's desired AWG size on the chart and trace it vertically. The number and size of strands needed to make the stranded conductor will be indicated by the horizontal line (strand number) and diagonal line (strand size) respectively.

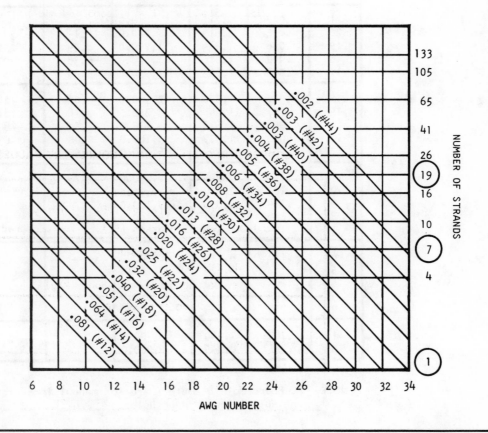

SOLID BARE WIRE GAGE DIAMETERS
AND PRINTED CIRCUIT DRILL SIZES

AMER WIRE GAGE (AWG)	NOM DIA	MM DIA	AREA CIRCULAR MILLS (sq. in.)	OHMS PER 1000 FT at 20°C	FEET PER POUND	PC BOARD HOLE DIA
36	.005	.127	25.0	415.	13,210.	.016 (#78)
35	.0056	.142	31.5	331.	10,481.	.016 (#78)
34	.006	.152	39.8	261.	8,310.	.016 (#78)
33	.007	.178	50.1	206.	6,591.	.016 (#78)
32	.008	.208	63.2	162.	5,227.	.016 (#78)
31	.009	.228	79.7	131.	4,145.	.016 (#78)
30	.010	.254	101.0	104.	3,287.	.016 (#78)
29	.011	.280	127.0	81.2	2,607.	.016 (#78)
28	.013	.330	160.0	65.3	2,067.	.016 (#78)
27	.014	.356	202.0	51.4	1,639.	.026 (#71)
26	.016	.406	254.0	41.0	1,300.	.026 (#71)
25	.018	.457	320.0	32.4	1,031.	.026 (#71)
24	.020	.508	404.0	25.7	817.7	.031 (#68)
23	.023	.584	509.0	20.3	648.4	.031 (#68)
22	.025	.635	642.0	16.2	514.2	.031 (#68)
21	.028	.711	810.0	12.8	407.8	.040 (#60)
20	.032	.813	1,020.0	10.1	323.4	.040 (#60)
19	.036	.914	1,290.0	8.04	256.5	.052 (#55)
18	.040	1.016	1,620.0	6.38	203.4	.052 (#55)
17	.045	1.143	2,050.0	5.04	161.3	.052 (#55)
16	.050	1.270	2,580.0	4.01	127.9	
15	.057	1.448	3,260.0	3.18	101.4	
14	.064	1.626	4,110.0	2.52	80.44	
13	.072	1.829	5,180.0	2.00	63.80	
12	.081	2.057	6,530.0	1.59	50.59	
11	.090	2.286	8,230.0	1.26	40.12	
10	.102	2.590	10,400.0	.999	31.82	
9	.114	2.896	13,100.0	.792	25.23	
8	.128	3.251	16,500.0	.628	20.01	
7	.144	3.658	20,800.0			
6	.162	4.115	26,300.0			
5	.182	4.623	33,100.0			
4	.204	5.182	41,700.0			
3	.229	5.817	52,600.0			
2	.258	6.553	66,400.0			
1	.289	7.340	83,700.0			
0	.325	8.255	106,000.0			

NOTE: PC Board Hole Diameters are for manual insertion and NOT plated through holes.

DESIGN REFERENCE MATERIALS

DRILL SIZES—DECIMAL AND METRIC EQUIVALENTS

SIZE	DECIMAL	MM	SIZE	DECIMAL	MM
85	.011	.2794	41	.096	2.4384
84	.0115	.2921	40	.098	2.4892
83	.012	.3048	39	.0995	2.5273
82	.0125	.3175	38	.1015	2.5780
81	.013	.3302	37	.104	2.6415
80	.0135	.3429	36	.1065	2.7050
79	.0145	.3683	7/64	.1094	2.7787
1/64	.0156	.3962	35	.110	2.7939
78	.016	.4064	34	.111	2.8193
77	.018	.4572	33	.113	2.8701
76	.020	.5080	32	.116	2.9463
75	.021	.5334	31	.120	3.0480
74	.0225	.5715	1/8	.125	3.1749
73	.024	.6096	30	.1285	3.2638
72	.025	.6350	29	.136	3.4543
71	.026	.6604	28	.1405	3.5686
70	.028	.7112	9/64	.1406	3.5712
69	.0292	.7417	27	.144	3.6575
68	.031	.7874	26	.147	3.7337
1/32	.0312	.7925	25	.1495	3.7972
67	.032	.8128	24	.152	3.8607
66	.033	.8382	23	.154	3.9115
65	.035	.8890	5/32	.1562	3.9674
64	.036	.9144	22	.157	3.9877
63	.037	.9398	21	.159	4.0385
62	.038	.9652	20	.161	4.0893
61	.039	.9906	19	.166	4.2163
60	.040	1.0160	18	.1695	4.3052
59	.041	1.0414	11/64	.1719	4.3662
58	.042	1.0668	17	.173	4.3941
57	.043	1.0922	16	.177	4.4957
56	.0465	1.1811	15	.180	4.5719
3/64	.0469	1.1912	14	.182	4.6227
55	.052	1.3208	13	.185	4.6989
54	.055	1.3970	3/16	.1875	4.7624
53	.0595	1.5113	12	.189	4.8005
1/16	.0625	1.5875	11	.191	4.8513
52	.0635	1.6129	10	.1935	4.9148
51	.067	1.7018	9	.196	4.9783
50	.070	1.7780	8	.199	5.0545
49	.073	1.8542	7	.201	5.1053
48	.076	1.9304	13/64	.2031	5.1586
5/64	.0781	1.9837	6	.204	5.1815
47	.0785	1.9939	5	.2055	5.2196
46	.081	2.0574	4	.209	5.3085
45	.082	2.0828	3	.213	5.4101
44	.086	2.1844	7/32	.2188	5.5574
43	.089	2.2606	2	.221	5.6133
42	.0935	2.3749	1	.228	5.7911
3/32	.0938	2.3825			

DESIGN REFERENCE MATERIALS

DRILL SIZES—DECIMAL AND METRIC EQUIVALENTS
(Continued)

SIZE	DECIMAL	MM	SIZE	DECIMAL	MM
A	.234	5.943	27/64	.4219	10.716
15/64	.2344	5.954	7/16	.4375	11.112
B	.238	6.045	29/64	.4531	11.509
C	.242	6.147	15/32	.4688	11.907
D	.246	6.248	31/64	.4844	12.304
E & ¼	.250	6.350	1/2	.5000	12.700
F	.257	6.528	33/64	.5156	13.096
G	.261	6.629	17/32	.5312	13.492
17/64	.2656	6.746	35/64	.5469	13.891
H	.266	6.756	9/16	.5625	14.287
I	.272	6.909	37/64	.5781	14.683
J	.277	7.036	19/32	.5938	15.082
K	.281	7.137	39/64	.6094	15.478
9/32	.2812	7.142	5/8	.625	15.875
L	.290	7.366	41/64	.6406	16.271
M	.295	7.493	21/32	.6562	16.667
19/64	.2969	7.541	43/64	.6719	17.066
N	.302	7.671	11/16	.6875	17.462
5/16	.3125	7.937	45/64	.7031	17.858
O	.316	8.026	23/32	.7188	18.257
P	.323	8.204	47/64	.7344	18.653
21/64	.3281	8.334	3/4	.750	19.050
Q	.332	8.433	49/64	.7656	19.446
R	.339	8.610	25/32	.7812	19.842
11/32	.3438	8.732	51/64	.7969	20.241
S	.348	8.839	13/16	.8125	20.637
T	.358	9.093	53/64	.8281	21.033
23/64	.3594	9.129	27/32	.8438	21.432
U	.368	9.347	55/64	.8594	21.828
3/8	.375	9.525	7/8	.875	22.225
V	.377	9.576	57/64	.8906	22.621
W	.386	9.804	29/32	.9062	23.017
25/64	.3906	9.921	59/64	.9219	23.416
X	.397	10.084	15/16	.9375	23.812
Y	.404	10.261	61/64	.9531	24.208
13/32	.4062	10.317	31/32	.9688	24.607
Z	.413	10.490	63/64	.9844	25.003
			1	1.000	25.400

NOTES: METRIC EQUIVALENTS ARE BASED
UPON 1 INCH = 25.3995mm.

TO FIND CENTIMETERS, DIVIDE
MILLIMETERS BY 10.

DESIGN REFERENCE MATERIALS

DECADE TABLE

VALUE	PREFIX	SYMBOL	EXAMPLE
$1\ 000\ 000\ 000\ 000 = 10^{12}$	tera	T	$THz = 10^{12}\,Hz$
$1\ 000\ 000\ 000 = 10^{9}$	giga	G	$GHz = 10^{9}\,Hz$
$1\ 000\ 000 = 10^{6}$	mega	M	$MHz = 10^{6}\,Hz$
$1\ 000 = 10^{3}$	kilo	k	$kV = 10^{3}\,V$
$100 = 10^{2}$	hecto	h	$hm = 10^{2}\,m$
$10 = 10$	deka	da	$dam = 10\,m$
$0.1 = 10^{-1}$	deci	d	$dm = 10^{-1}\,m$
$0.01 = 10^{-2}$	centi	c	$cm = 10^{-2}\,m$
$0.001 = 10^{-3}$	milli	m	$mA = 10^{-3}\,A$
$0.000\ 001 = 10^{-6}$	micro	μ	$\mu V = 10^{-6}\,V$
$0.000\ 000\ 001 = 10^{-9}$	nano	n	$ns - 10^{-9}\,s$
$0.000\ 000\ 000\ 001 = 10^{-12}$	pico	p	$pF = 10^{-12}\,F$

CONDUCTOR THICKNESS AND WIDTH

(For use in determining current carrying capacity and sizes of etched copper conductors for various temperature rises above ambient)

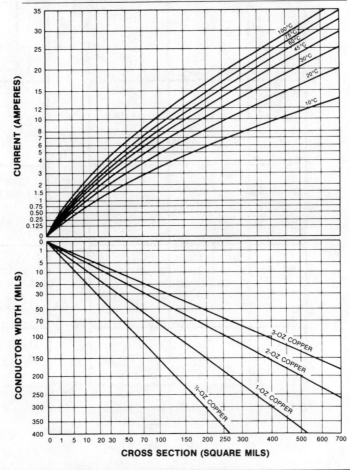

DESIGN REFERENCE MATERIALS

TEMPERATURE CONVERSION

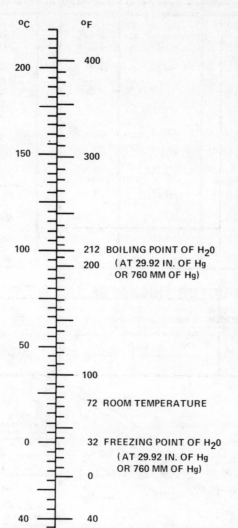

°C	°F	°C	°F
−70	−94	110	230
−60	−76	120	248
−50	−58	130	266
−40	−40	140	284
−30	−22	150	302
−20	− 4	160	320
−10	14	170	338
0	32	180	356
10	50	190	374
20	68	200	392
30	86	210	410
40	104	220	428
50	122	230	446
60	140	240	464
70	158	250	482
80	176	260	500
100	212	300	572

212 BOILING POINT OF H_2O
(AT 29.92 IN. OF Hg
OR 760 MM OF Hg)

72 ROOM TEMPERATURE

32 FREEZING POINT OF H_2O
(AT 29.92 IN. OF Hg
OR 760 MM OF Hg)

VISCOSITY INTERPRETATION

Listing viscosities by centipoise is sometimes confusing unless one is thoroughly familiar with this particular system. The following practical interpretation will be of assistance. The viscosity readings are for mixed compounds.

The readings were taken at $25°C$ with a Brookfield viscosimeter.

```
        1 CENTIPOISE (C.P.S.) - WATER
      400 CENTIPOISE (C.P.S.) - #10 MOTOR OIL
    1,000 CENTIPOISE (C.P.S.) - CASTOR OIL
    3,500 CENTIPOISE (C.P.S.) - KARO SYRUP
    4,500 CENTIPOISE (C.P.S.) - #40 MOTOR OIL
   25,000 CENTIPOISE (C.P.S.) - HERSHEY CHOCOLATE SYRUP
```

To convert temperatures
use following formulas:

$°C = 5/9 \ (°F - 32)$ (Celsius) Formerly Centigrade

$°F = 9/5 \ (°C + 32)$ (Fahrenheit)

$°K = °C + 273.15$ (Kelvin)

$°R = °F + 459.67$ (Rankin)

CONVERSION TABLES

1 hp = 746 watts = 33,000 ft-lb/min = 2,544 btu/hr	1 in of water = .0361 lb per sq in = .0735 in of Hg	1 kw = 1,000 watts = 1.34 hp = 44,240 ft-lb/min = 56.9 btu/min
1 hp-hr = .746 kw-hr = 1,980,000 ft-lb = 2,545 btu = 273,740 kg meters	1 ft of water = .4332 lb per sq in = .8824 in of Hg 1 in of Hg = .4912 lb per sq in = 13.58 in of water = 1.131 ft of water	1 kw hr = 1,000 watt hr = 1.34 hp-hr = 2,654,200 ft-lb = 3,413 btu = 3,600,000 joules
1 ft-lb = 1,356 joules = 1,3826 kg meters	1 cm of Hg = .1934 lb per sq in 1 atm = 14,696 lb per sq in = 33.95 ft of water = 760 mm of Hg	1 btu = 1052 watt sec = 778 ft-lb
1 watt = 1 joule/sec = 3.413 btu/hr = 44.22 ft-lb/min	1 lb per sq in = 27.71 in of water = 2.309 ft of water = 2.04 in of Hg = .06804 atm	1 joule = 1 watt sec = .73756 ft-lb
1 inch = 25.4 min 1 mm = .03937 in. 1 foot = 30.48 cm 1 meter = 39.37 in. = 3.28 ft 1 meter = 100 cm = 1000 mm 1 mile = 1.609 km 1 km = .6214 mile = 3/5 mile = 1000 meters	1 sq in = 6.4516 sq cm 1 sq cm = .155 sq in 1 sq ft = 929.03 sq cm	1 kg = 2.2 lb 1 cu in = 16.39 cu cm 1 cu ft = 1728 cu in 1 cu ft = 7.4805 US gal 1 US gal = .1337 cu ft 1 liter = 61.0 cu in = 1000 cu cm (cc) 1 liter = 1.0567 US qt
1 mile = 1,760 yd = 5,280 ft 1 sq mi = 640 acres 1 cir in = 1,000,000 cir mils	1 nautical mile = 6,080 ft 1 fathom = 6 ft	1 cu ft = 1.728 cu in 1 ounce = 437.5 grams = 28.35 grams

SURFACES AND VOLUME OF SOLIDS

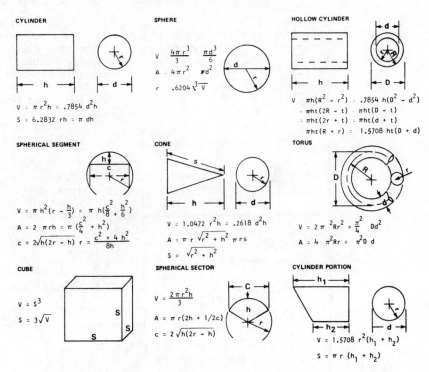

CYLINDER

$V = \pi r^2 h = .7854\, d^2 h$
$S = 6.2832\, rh = \pi\, dh$

SPHERE

$V\quad \dfrac{4\pi r^3}{3}\quad \dfrac{\pi d^3}{6}$
$A\quad 4\pi r^2\quad \pi d^2$
$r\quad .6204 \sqrt[3]{V}$

HOLLOW CYLINDER

$V \quad \pi h(R^2 - r^2) = .7854\, h(D^2 - d^2)$
$= \pi ht(2R - t) = \pi ht(D - t)$
$= \pi ht(2r + t) = \pi ht(d + t)$
$\pi ht(R + r) = 1.5708\, ht(D + d)$

SPHERICAL SEGMENT

$V = \pi h^2(r - \dfrac{h}{3}) = \pi\, h(\dfrac{c^2}{8} + \dfrac{h^2}{6})$
$A = 2\, \pi rh = \pi\,(\dfrac{c^2}{4} + h^2)$
$c = 2\sqrt{h(2r - h)} \quad r = \dfrac{c^2 + 4h^2}{8h}$

CONE

$V = 1.0472\, r^2 h = .2618\, d^2 h$
$A = \pi r \sqrt{r^2 + h^2}\ \ \pi rs$
$S = \sqrt{r^2 + h^2}$

TORUS

$V = 2\, \pi^2 Rr = \dfrac{\pi^2}{4}\, Dd^2$
$A = 4\ \pi^2 Rr = \pi^2 D\, d$

CUBE

$V = s^3$
$S = 3\sqrt{V}$

SPHERICAL SECTOR

$V = \dfrac{2\, \pi r^2 h}{3}$
$A = \pi r(2h + 1/2c)$
$c = 2\sqrt{h(2r - h)}$

CYLINDER PORTION

$V = 1.5708\, r^2(h_1 + h_2)$
$S = \pi r\, (h_1 + h_2)$

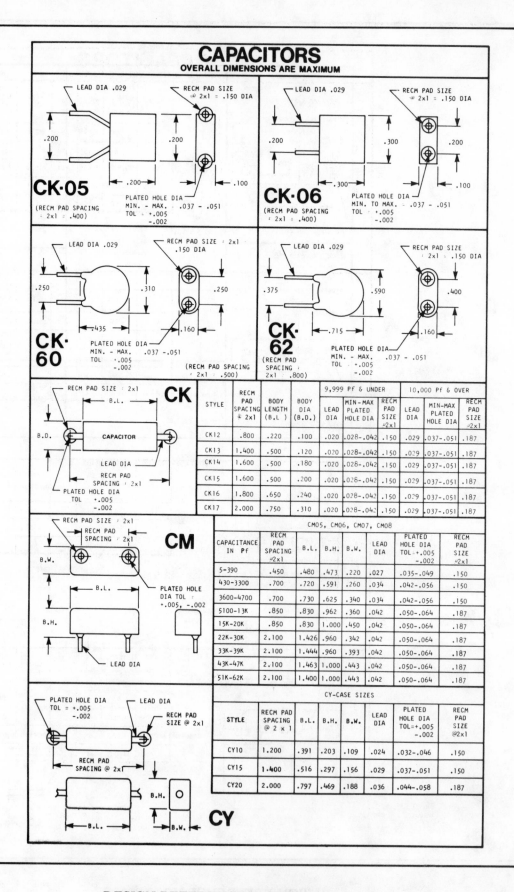

CAPACITORS
OVERALL DIMENSIONS ARE MAXIMUM

CK-05 — LEAD DIA .029; RECM PAD SIZE @ 2x1 = .150 DIA; .200; .200; .200; .100; PLATED HOLE DIA MIN.-MAX. .037-.051 TOL = +.005 -.002 (RECM PAD SPACING = 2x1 = .400)

CK-06 — LEAD DIA .029; RECM PAD SIZE @ 2x1 = .150 DIA; .200; .300; .200; .300; .100; PLATED HOLE DIA MIN. TO MAX. = .037-.051 TOL = +.005 -.002 (RECM PAD SPACING = 2x1 = .400)

CK-60 — LEAD DIA .029; RECM PAD SIZE = 2x1 .150 DIA; .250; .310; .250; .435; .160; PLATED HOLE DIA MIN.-MAX. .037-.051 TOL +.005 -.002 (RECM PAD SPACING = 2x1 = .500)

CK-62 — LEAD DIA .029; RECM PAD SIZE = 2x1 = .150 DIA; .375; .590; .400; .715; .160; PLATED HOLE DIA MIN.-MAX. .037-.051 TOL = +.005 -.002 (RECM PAD SPACING = 2x1 .800)

CK — RECM PAD SIZE = 2x1; B.L.; B.D.; CAPACITOR; LEAD DIA; RECM PAD SPACING = 2x1; PLATED HOLE DIA TOL +.005 -.002

STYLE	RECM PAD SPACING @ 2x1	BODY LENGTH (B.L.)	BODY DIA (B.D.)	9,999 Pf & UNDER			10,000 Pf & OVER		
				LEAD DIA	MIN-MAX PLATED HOLE DIA	RECM PAD SIZE 2x1	LEAD DIA	MIN-MAX PLATED HOLE DIA	RECM PAD SIZE 2x1
CK12	.800	.220	.100	.020	.028-.042	.150	.029	.037-.051	.187
CK13	1.400	.500	.120	.020	.028-.042	.150	.029	.037-.051	.187
CK14	1.600	.500	.180	.020	.028-.042	.150	.029	.037-.051	.187
CK15	1.600	.500	.200	.020	.028-.042	.150	.029	.037-.051	.187
CK16	1.800	.650	.240	.020	.028-.042	.150	.029	.037-.051	.187
CK17	2.000	.750	.310	.020	.028-.042	.150	.029	.037-.051	.187

CM — RECM PAD SIZE = 2x1; RECM PAD SPACING = 2x1; B.W.; B.L.; B.H.; PLATED HOLE DIA TOL = +.005, -.002; LEAD DIA

CAPACITANCE IN Pf	RECM PAD SPACING 2x1	B.L.	B.H.	B.W.	LEAD DIA	PLATED HOLE DIA TOL = +.005 -.002	RECM PAD SIZE 2x1
CM05, CM06, CM07, CM08							
5-390	.450	.480	.473	.220	.027	.035-.049	.150
430-3300	.700	.720	.591	.260	.034	.042-.056	.150
3600-4700	.700	.730	.625	.340	.034	.042-.056	.150
5100-13K	.850	.830	.962	.360	.042	.050-.064	.187
15K-20K	.850	.830	1.000	.450	.042	.050-.064	.187
22K-30K	2.100	1.426	.960	.342	.042	.050-.064	.187
33K-39K	2.100	1.444	.960	.393	.042	.050-.064	.187
43K-47K	2.100	1.463	1.000	.443	.042	.050-.064	.187
51K-62K	2.100	1.400	1.000	.443	.042	.050-.064	.187

CY — PLATED HOLE DIA TOL = +.005 -.002; LEAD DIA; RECM PAD SIZE @ 2x1; RECM PAD SPACING @ 2x1; B.H.; B.L.; B.W.

STYLE	RECM PAD SPACING @ 2 x 1	B.L.	B.H.	B.W.	LEAD DIA	PLATED HOLE DIA TOL=+.005 -.002	RECM PAD SIZE @2x1
CY-CASE SIZES							
CY10	1.200	.391	.203	.109	.024	.032-.046	.150
CY15	1.400	.516	.297	.156	.029	.037-.051	.150
CY20	2.000	.797	.469	.188	.036	.044-.058	.187

CAPACITORS
OVERALL DIMENSIONS ARE MAXIMUM

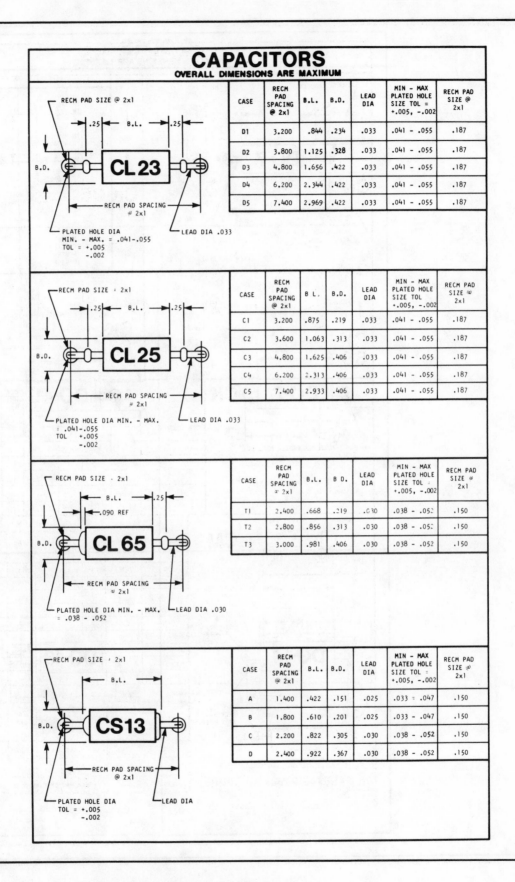

CASE	RECM PAD SPACING @ 2x1	B.L.	B.D.	LEAD DIA	MIN – MAX PLATED HOLE SIZE TOL = +.005, –.002	RECM PAD SIZE @ 2x1
D1	3.200	.844	.234	.033	.041 – .055	.187
D2	3.800	1.125	.328	.033	.041 – .055	.187
D3	4.800	1.656	.422	.033	.041 – .055	.187
D4	6.200	2.344	.422	.033	.041 – .055	.187
D5	7.400	2.969	.422	.033	.041 – .055	.187

CL 23

RECM PAD SIZE @ 2x1
.25 — B.L. — .25
B.D.
RECM PAD SPACING @ 2x1
PLATED HOLE DIA MIN. – MAX. = .041–.055 TOL = +.005 –.002
LEAD DIA .033

CASE	RECM PAD SPACING @ 2x1	B L.	B.D.	LEAD DIA	MIN – MAX PLATED HOLE SIZE TOL +.005, –.002	RECM PAD SIZE @ 2x1
C1	3.200	.875	.219	.033	.041 – .055	.187
C2	3.600	1.063	.313	.033	.041 – .055	.187
C3	4.800	1.625	.406	.033	.041 – .055	.187
C4	6.200	2.313	.406	.033	.041 – .055	.187
C5	7.400	2.933	.406	.033	.041 – .055	.187

CL 25

RECM PAD SIZE · 2x1
.25 — B.L. — .25
B.D.
RECM PAD SPACING @ 2x1
PLATED HOLE DIA MIN. – MAX. = .041–.055 TOL +.005 –.002
LEAD DIA .033

CASE	RECM PAD SPACING @ 2x1	B.L.	B D.	LEAD DIA	MIN – MAX PLATED HOLE SIZE TOL = +.005, –.002	RECM PAD SIZE @ 2x1
T1	2.400	.668	.219	.030	.038 – .052	.150
T2	2.800	.856	.313	.030	.038 – .052	.150
T3	3.000	.981	.406	.030	.038 – .052	.150

CL 65

RECM PAD SIZE · 2x1
B.L. — .25
.090 REF
B.D.
RECM PAD SPACING @ 2x1
PLATED HOLE DIA MIN. – MAX. = .038 – .052
LEAD DIA .030

CASE	RECM PAD SPACING @ 2x1	B.L.	B.D.	LEAD DIA	MIN – MAX PLATED HOLE SIZE TOL = +.005, –.002	RECM PAD SIZE @ 2x1
A	1.400	.422	.151	.025	.033 = .047	.150
B	1.800	.610	.201	.025	.033 – .047	.150
C	2.200	.822	.305	.030	.038 – .052	.150
D	2.400	.922	.367	.030	.038 – .052	.150

CS 13

RECM PAD SIZE · 2x1
B.L.
B.D.
RECM PAD SPACING @ 2x1
PLATED HOLE DIA TOL = +.005 –.002
LEAD DIA

DIODES
OVERALL DIMENSIONS ARE MAXIMUM

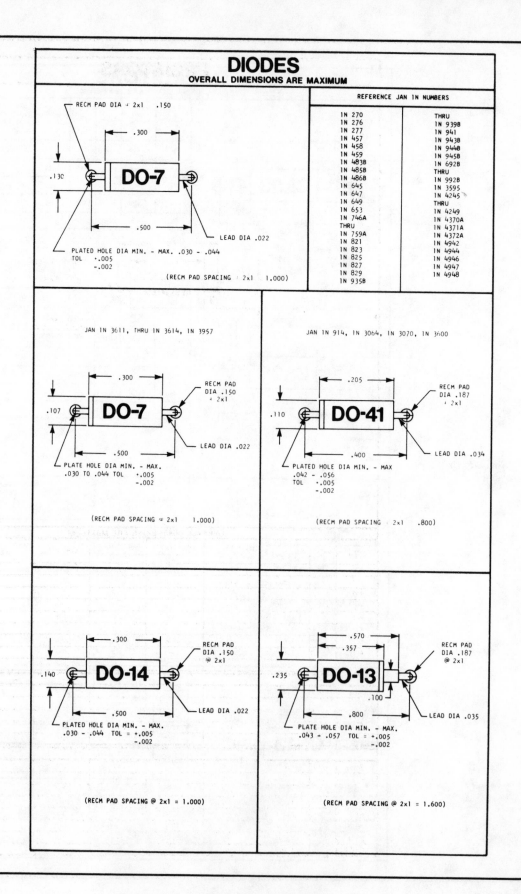

RECM PAD DIA = 2x1 .150

.300

.130

DO-7

.500

LEAD DIA .022

PLATED HOLE DIA MIN. - MAX. .030 - .044
TOL +.005
     -.002

(RECM PAD SPACING = 2x1 1.000)

REFERENCE JAN 1N NUMBERS	
1N 270	
1N 276	THRU
1N 277	1N 939B
1N 457	1N 941
1N 458	1N 943B
1N 459	1N 944B
1N 483B	1N 945B
1N 485B	1N 692B
1N 486B	THRU
1N 645	1N 992B
1N 647	1N 3595
1N 649	1N 4245
1N 653	THRU
1N 746A	1N 4249
THRU	1N 4370A
1N 759A	1N 4371A
1N 821	1N 4372A
1N 823	1N 4942
1N 825	1N 4944
1N 827	1N 4946
1N 829	1N 4947
1N 935B	1N 4948

JAN 1N 3611, THRU 1N 3614, 1N 3957

.300

.107

DO-7

RECM PAD
DIA .150
= 2x1

.500

LEAD DIA .022

PLATE HOLE DIA MIN. - MAX.
.030 TO .044 TOL +.005
                  -.002

(RECM PAD SPACING @ 2x1   1.000)

JAN 1N 914, 1N 3064, 1N 3070, 1N 3600

.205

.110

DO-41

RECM PAD
DIA .187
= 2x1

.400

LEAD DIA .034

PLATED HOLE DIA MIN. - MAX
.042 - .056
TOL +.005
     -.002

(RECM PAD SPACING = 2x1   .800)

.300

.140

DO-14

RECM PAD
DIA .150
@ 2x1

.500

LEAD DIA .022

PLATED HOLE DIA MIN. - MAX.
.030 - .044 TOL = +.005
                 -.002

(RECM PAD SPACING @ 2x1 = 1.000)

.570

.357

.235

DO-13

RECM PAD
DIA .187
@ 2x1

.100

.800

LEAD DIA .035

PLATE HOLE DIA MIN. - MAX.
.043 - .057 TOL = +.005
                   -.002

(RECM PAD SPACING @ 2x1 = 1.600)

DESIGN REFERENCE MATERIALS

RESISTORS

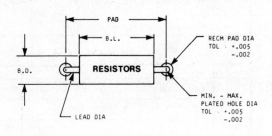

OVERALL DIMENSIONS ARE MAXIMUM

STYLE	WATT	PAD · 2 × 1	RECM PAD DIA · 2 × 1	BODY LENGTH (B.L.)	BODY DIA (B.D.)	LEAD DIA	MIN. – MAX. PLATED HOLE DIA TOL = +.005 / –.002
RC05	1/8	.800	.150	.160	.066	.018	.026 – .040
RC06	1/10	.800	.150	.160	.075	.018	.026 – .040
RC07	1/4	1.000	.150	.281	.098	.027	.035 – .049
RC08	1/10	.800	.150	.225	.115	.018	.026 – .040
RC12	1/4	1.200	.150	.390	.148	.028	.036 – .050
RC20	1/2	1.200	.187	.416	.161	.036	.044 – .058
RC22	1/2	1.600	.187	.575	.233	.036	.044 – .058
RC32	1	1.600	.187	.593	.240	.042	.050 – .064
RC42	2	2.000	.200	.728	.336	.048	.056 – .070
RL05	1/8	.800	.150	.170	.070	.017	.025 – .040
RL07	1/4	1.000	.150	.281	.098	.027	.035 – .049
RL20	1/2	1.200	.187	.416	.161	.036	.044 – .058
RL32	1	1.600	.187	.593	.205	.042	.050 – .064
RL42	2	2.000	.200	.728	.336	.050	.058 – .072
RN50	1/10	.800	.150	.170	.080	.018	.026 – .040
RN55	1/10	1.000	.150	.281	.141	.027	.035 – .049
RN60	1/8	1.200	.150	.437	.165	.027	.035 – .049
RN65	1/4	1.800	.150	.656	.250	.027	.035 – .049
RN70	1/2	2.200	.187	.875	.328	.034	.042 – .056
RN75	1	2.800	.187	1.125	.437	.034	.042 – .056
RN80	2	5.000	.187	2.281	.437	.034	.042 – .056
RWP20	3	1.600	.187	.563	.250	.036	.044 – .058

TRANSISTORS
OVERALL DIMENSIONS ARE MAXIMUM

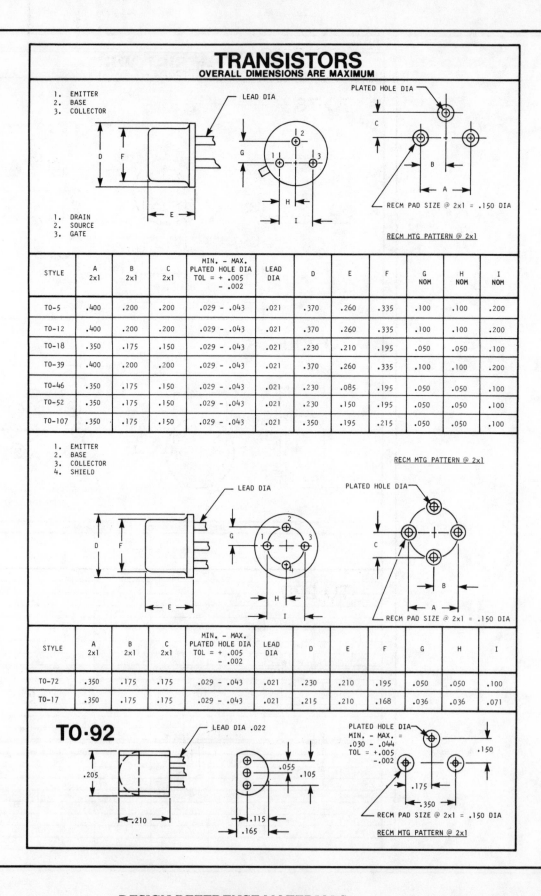

1. EMITTER
2. BASE
3. COLLECTOR

LEAD DIA

PLATED HOLE DIA

RECM PAD SIZE @ 2x1 = .150 DIA

RECM MTG PATTERN @ 2x1

1. DRAIN
2. SOURCE
3. GATE

STYLE	A 2x1	B 2x1	C 2x1	MIN. – MAX. PLATED HOLE DIA TOL = + .005 – .002	LEAD DIA	D	E	F	G NOM	H NOM	I NOM
TO-5	.400	.200	.200	.029 – .043	.021	.370	.260	.335	.100	.100	.200
TO-12	.400	.200	.200	.029 – .043	.021	.370	.260	.335	.100	.100	.200
TO-18	.350	.175	.150	.029 – .043	.021	.230	.210	.195	.050	.050	.100
TO-39	.400	.200	.200	.029 – .043	.021	.370	.260	.335	.100	.100	.200
TO-46	.350	.175	.150	.029 – .043	.021	.230	.085	.195	.050	.050	.100
TO-52	.350	.175	.150	.029 – .043	.021	.230	.150	.195	.050	.050	.100
TO-107	.350	.175	.150	.029 – .043	.021	.350	.195	.215	.050	.050	.100

1. EMITTER
2. BASE
3. COLLECTOR
4. SHIELD

RECM MTG PATTERN @ 2x1

LEAD DIA

PLATED HOLE DIA

RECM PAD SIZE @ 2x1 = .150 DIA

STYLE	A 2x1	B 2x1	C 2x1	MIN. – MAX. PLATED HOLE DIA TOL = + .005 – .002	LEAD DIA	D	E	F	G	H	I
TO-72	.350	.175	.175	.029 – .043	.021	.230	.210	.195	.050	.050	.100
TO-17	.350	.175	.175	.029 – .043	.021	.215	.210	.168	.036	.036	.071

TO·92

LEAD DIA .022

.205

.210

.055
.105
.115
.165

PLATED HOLE DIA
MIN. – MAX. =
.030 – .044
TOL = +.005
–.002

.150

.175

.350

RECM PAD SIZE @ 2x1 = .150 DIA

RECM MTG PATTERN @ 2x1

DESIGN REFERENCE MATERIALS

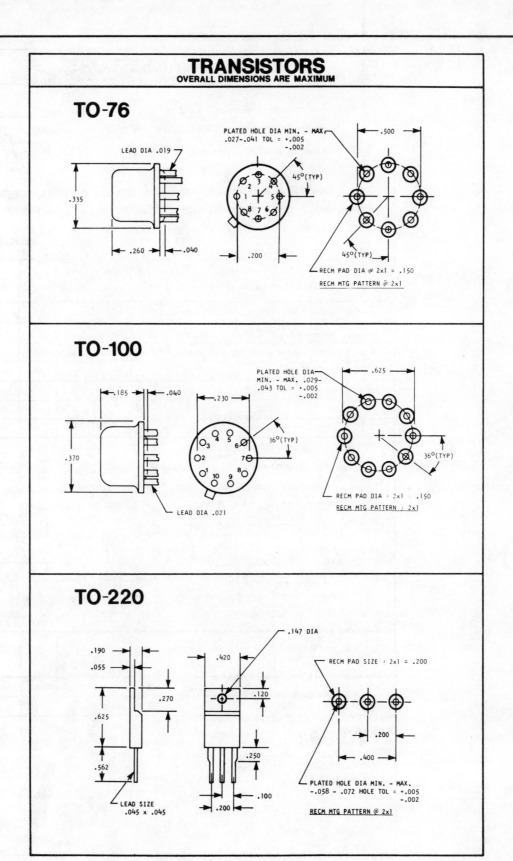

TRANSISTORS
OVERALL DIMENSIONS ARE MAXIMUM

TO-76

LEAD DIA .019

.335

.260   .040

PLATED HOLE DIA MIN. - MAX.
.027-.041 TOL = +.005
                -.002

45°(TYP)

.200

.500

45°(TYP)

RECM PAD DIA @ 2x1 = .150
RECM MTG PATTERN @ 2x1

TO-100

.185   .040

.230

.370

LEAD DIA .021

PLATED HOLE DIA
MIN. - MAX. .029-
.043 TOL = +.005
              -.002

.625

36°(TYP)

36°(TYP)

RECM PAD DIA : 2x1 . .150
RECM MTG PATTERN : 2x1

TO-220

.190
.055

.270

.625

.562

LEAD SIZE
.045 x .045

.420

.147 DIA

.120

.250

.100

.200

RECM PAD SIZE : 2x1 = .200

.200

.400

PLATED HOLE DIA MIN. - MAX.
-.058 - .072 HOLE TOL = +.005
                            -.002

RECM MTG PATTERN @ 2x1

DESIGN REFERENCE MATERIALS

FUNDAMENTAL RULES FOR DIMENSIONING

1. Show enough dimensions so that the intended sizes and shapes can be determined without calculating or assuming any distances.

2. State each dimension clearly so that it can be interpreted in only one way.

3. SHOW THE DIMENSIONS BETWEEN POINTS, LINES, OR SURFACES WHICH HAVE A NECESSARY AND SPECIFIC RELATION TO EACH OTHER OR WHICH CONTROL THE LOCATION OF OTHER COMPONENTS OR MATING PARTS.

4. Select and arrange dimensions to avoid accumulations of tolerances that may permit various interpretations and cause unsatisfactory mating of parts and failure in use.

5. Show each dimension only once.

6. Where possible, dimension each feature in the view where it appears in profile and where its true shape is evident.

7. Do not show dimensions to lines representing hidden surfaces.

APPROXIMATE SURFACE ROUGHNESS

USE OF SURFACE ROUGHNESS SYMBOLS

The roughness of any surface, except cast surfaces, may be expressed by the use of symbols and numerical ratings. The finish mark symbols are used to designate the applicable surface. The numerical rating indicates the roughness of that surface.

CONTROL OF SURFACE ROUGHNESS

The roughness of any surface is determined by design requirements rather than by production processes; however, overcontrol will increase the cost of production. Therefore, specify surface roughness control on drawings only when it is essential to the appearance or mechanical performance of the product.

TYPICAL SURFACE ROUGHNESS CHART

Castings: Do not use symbols and ratings on cast surfaces.
Mill Cleanup and Shear Cuts..250 to 125
Machining:

    .001 to .005 Tolerance. 63
    .0005 to .001 Tolerance 32
    .0002 to .0005 Tolerance. 16
    .0000 to .0002 Tolerance. 8

FINISH MARK SYMBOL AND ROUGHNESS HEIGHT RATING

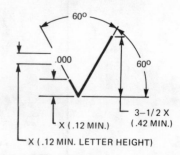

CLEARANCE HOLES FOR THREADED FASTENERS

THREAD SIZE	TOLERANCE BETWEEN HOLE CENTERS		
	±.005	±.010	±.020
0	.076 DIA (#48)		
1	.089 DIA (#43)	.102 DIA (#38)	.128 DIA (#30)*
2	.102 DIA (#38)	.116 DIA (#32)	.144 DIA (#27)*
4	.125 DIA (#1/8)	.140 DIA (#28)	.166 DIA (#19)*
6	.150 DIA (#25)	.166 DIA (#19)	.191 DIA (#11)*
8	.177 DIA (#16)	.189 DIA (#12)	.219 DIA (#7/32)*
10	.199 DIA (#8)	.213 DIA (#3)	.242 DIA (C)
1/4	.257 DIA (F)	.272 DIA (I)	.302 DIA (N)

*Exclude Fillister and socket head cap screws.
FLAT WASHERS WILL BE REQUIRED WITH ±.020 TOLERANCE.

NOTES:

1. This chart applies to the following:

   a. Where a row of holes is on the same centerline.

   b. Where the holes are on more than one centerline.

   c. Where clearance holes are in both parts.

   d. Where clearance holes are in one part and tapped holes or studs are in the other part.

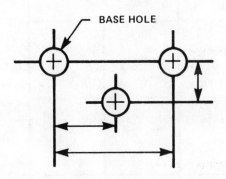

All dimensioning shall be from an established base hole or fixed point.

2. This chart does not apply to the following:

   a. Clearance holes for rivets.

   b. One fastener used independent of any other.

   c. Clearance holes for flat head machine screws.

3. The hole sizes listed have been formulated for the most severe conditions for screw fasteners and represent a noninterference probability of approximately 98%.

PREFERRED BEND RADII FOR STRAIGHT BENDS IN SHEET METALS

SHEET THICKNESS	ALUMINUM						
	2024 T3	5052 0	5052 H32	5052 H34	6061 0	6061 T4	6061 T6 (See Note 1)
.020	.06	.03	.03	.03*	.03	.06	.06
.025	.06	.03	.03	.03*	.03*	.06	.06
.032	.09*	.03	.06	.06*	.03*	.09	.09*
.040	.12	.06*	.06	.06*	.06	.12	.12*
.050	.16	.06	.09	.09*	.06*	.16	.16*
.063	.19*	.06	.09*	.09*	.06*	.19	.19*
.080	.25	.09	.12*	.12	.09	.25	.25*
.090	.31*	.09	.12*	.12	.09	.31	.31*
.100	.38	.12	.19	.19	.12	.38	.38
.125	.50*	.12	.19*	.19	.12	.50	.50*
.160	.75	.16	.25	.25	.16	.62	.62
.190	1.00*	.19	.37*	.37	.19	.84	.87*

SHEET THICKNESS	STEEL		PLAIN CARBON
	CORROSION RESISTANT		
	TYPES 301 302 304 (Annealed)	TYPES 301 302 304 (1/4 H)	
.020			.06
.025			.06
.032			.06
.040			.09
.050			.09
.063			.09
.080			.12
.090			.12
.100			.16
.125			.19
.160			.25
.190			.31
.020-.040	.03	.06	
.045-.070	.06	.12	
.075-.105	.09	.19	
.110-.135	.12	.25	

NOTES:

1. If 6061-T6 material is used and the minimum bend radii shown is too large, the option is to anneal the material, form the bend, and then heat treat the material to the T6 condition again.

2. *Represents "common stock" material.

3. Shops require .010 tolerance for each bend.

DESIGN REFERENCE MATERIALS

DISSIMILAR METALS

Without proper protective coating, do not use a metal from
one group in contact with a metal listed in another group:

1. Tin, cadmium and zinc may be used with all metals
   in Groups II and III.

2. Stainless steel may be used with all metals in
   Groups II, III and IV.

Use metals within each group together.
MIL-STD-454

GROUP I	GROUP II	GROUP III	GROUP IV
MAGNESIUM ALLOYS	ALUMINUM	ZINC	COPPER AND ITS ALLOYS
TIN	ALUMINUM ALLOYS	CADMIUM	NICKEL AND ITS ALLOYS
Al ALLOY (5052)	ZINC	STEEL	CHROMIUM
Al ALLOY (5056)	CADMIUM	LEAD	STAINLESS STEEL
Al ALLOY (5356)	TIN	TIN	GOLD
		TIN LEAD (SOLDER)	
Al ALLOY (6061)	STAINLESS STEEL	STAINLESS STEEL	SILVER
Al ALLOY (6063)	TIN LEAD (SOLDER)	NICKEL AND ITS ALLOYS	

KNURL DATA

For decoration or gripping application, specify pitch and
type. Add "Before" and "After Knurling" dimensions to call-
out when application is for a press fit between two parts.

REF: USAS Y14.5-1966, USAS B94.6-1966

TYPE	DIAMETRAL PITCH	INCREASE IN PART DIAMETER WHEN KNURLED (APPROX.)
DIAMOND	160 DP (Fine) 128 DP (Med) 96 DP (Coarse)*	.009 .012 .016
DIAGONAL (PLAIN SPIRAL)	160 DP (Fine) 128 DP (Med) 96 DP (Coarse)*	.009 .012 .016
STRAIGHT (90° Vee)	160 DP (Fine) 128 DP (Med) 96 DP (Coarse)*	.009 .012 .016

* Preferred Diametral Pitch

RIVETS

RIVET DIA	PILOT DRILL	BODY DRILL	100° CSK RIVET		UNIVERSAL HD
			±.004 HEAD DIA	CSK DIA	HEAD DIA
1/16	-	.067 (#51)	.114	.114	.125
3/32	-	.098 (#40)	.179	.179	.187
1/8	.098 (#40)	.128 (#30)	.225	.225	.250
5/32	.098 (#40)	.159 (#21)	.286	.286	.312
3/16	.098 (#40)	.191 (#11)	.353	.353	.375
1/4	.098 (#40)	.257 (F)	.476	.476	.500

NOTES:

1. Mating parts to be riveted in assembly should have all predrilled pilot holes in one of the mating parts. Selection of part to be predrilled may be determined by:

    a. Accessibility for hand drilling.

    b. The part that has the greater material thickness.

    c. The material of greater hardness.

2. Pilot holes for rivets 1/8 diameter or more shall be .098 Dia (#40).

3. Rivet Diameter:

    Maximum - 3 times thickness of material to be riveted.

    Minimum - Thickness of material to be riveted.

4. Minimum Edge Distance from Center of Rivet:

    2 times diameter of rivet.

5. SELECTING RIVET LENGTH

    SUM OF MATERIAL THICKNESS PLUS 1-1/2 TIMES DIAMETER OF RIVET (FOR UPSETTING HEAD); TO NEAREST 1/16-INCH INCREMENT SHORTER THAN MAXIMUM CALCULATED LENGTH.

6. Types of Commonly Used Rivets

    UNIVERSAL HD (Aluminum) - TI 411175; MS 20470 (Preferred usage)

    100° CSK HD (Aluminum) - TI 411172; MS 20426

    UNIVERSAL HD (Steel, Carbon, and Steel Corrosion-Resistant) - TI 416918; MS 20613

    UNIVERSAL HD (Brass, Copper and Nickel-Copper Alloy) - TI 418180; MS 20615

    Tubular, Oval HD (Aluminum, Steel, Brass, Monel) - TI 411059; MS 16535

DESIGN REFERENCE MATERIALS

O·80

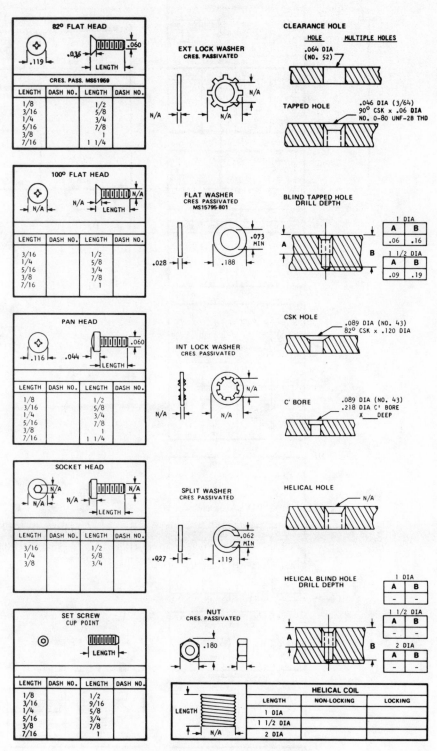

82° FLAT HEAD

.119 .036 .060 LENGTH

CRES. PASS. MS51969

LENGTH	DASH NO.	LENGTH	DASH NO.
1/8		1/2	
3/16		5/8	
1/4		3/4	
5/16		7/8	
3/8		1	
7/16		1 1/4	

100° FLAT HEAD

N/A N/A N/A LENGTH

LENGTH	DASH NO.	LENGTH	DASH NO.
3/16		1/2	
1/4		5/8	
5/16		3/4	
3/8		7/8	
7/16		1	

PAN HEAD

.116 .044 .060 LENGTH

LENGTH	DASH NO.	LENGTH	DASH NO.
1/8		1/2	
3/16		5/8	
1/4		3/4	
5/16		7/8	
3/8		1	
7/16		1 1/4	

SOCKET HEAD

N/A N/A N/A LENGTH

LENGTH	DASH NO.	LENGTH	DASH NO.
3/16		1/2	
1/4		5/8	
3/8		3/4	

SET SCREW CUP POINT

LENGTH

LENGTH	DASH NO.	LENGTH	DASH NO.
1/8		1/2	
3/16		9/16	
1/4		5/8	
5/16		3/4	
3/8		7/8	
7/16		1	

EXT LOCK WASHER CRES. PASSIVATED — N/A N/A

FLAT WASHER CRES PASSIVATED MS15795-801 — .028 .188 .073 MIN

INT LOCK WASHER CRES. PASSIVATED — N/A N/A

SPLIT WASHER CRES PASSIVATED — .027 .119 .062 MIN

NUT CRES. PASSIVATED — .180

CLEARANCE HOLE

HOLE   MULTIPLE HOLES

.064 DIA (NO. 52)

TAPPED HOLE — .046 DIA (3/64) 90° CSK x .06 DIA NO. 0-80 UNF-2B THD

BLIND TAPPED HOLE DRILL DEPTH

1 DIA	
A	B
.06	.16

1 1/2 DIA	
A	B
.09	.19

CSK HOLE — .089 DIA (NO. 43) 82° CSK x .120 DIA

C' BORE — .089 DIA (NO. 43) .218 DIA C' BORE X____ DEEP

HELICAL HOLE — N/A

HELICAL BLIND HOLE DRILL DEPTH

1 DIA	
A	B
-	-

1 1/2 DIA	
A	B
-	-

2 DIA	
A	B
-	-

HELICAL COIL

LENGTH	NON-LOCKING	LOCKING
1 DIA		
1 1/2 DIA		
2 DIA		

ALL DIMENSIONS MAXIMUM UNLESS OTHERWISE SPECIFIED

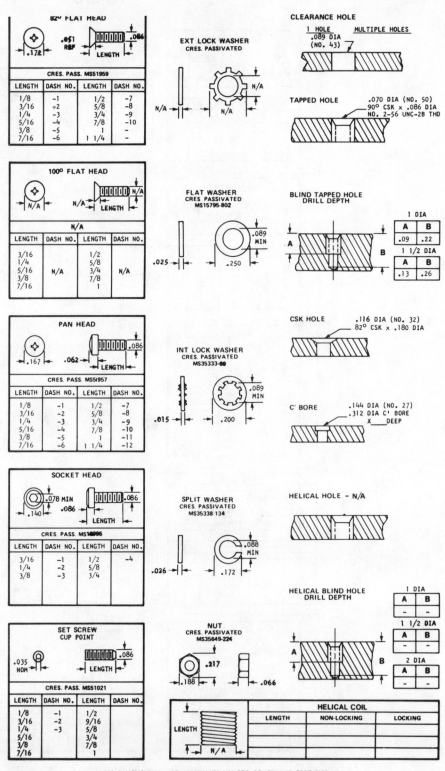

82° FLAT HEAD

.172 | .051 REF | .086 | LENGTH

CRES. PASS. MS51959

LENGTH	DASH NO.	LENGTH	DASH NO.
1/8	-1	1/2	-7
3/16	-2	5/8	-8
1/4	-3	3/4	-9
5/16	-4	7/8	-10
3/8	-5	1	-
7/16	-6	1 1/4	-

100° FLAT HEAD

N/A | LENGTH | N/A

N/A

LENGTH	DASH NO.	LENGTH	DASH NO.
3/16		1/2	
1/4		5/8	
5/16	N/A	3/4	N/A
3/8		7/8	
7/16		1	

PAN HEAD

.167 | .062 | .086 | LENGTH

CRES. PASS. MS51957

LENGTH	DASH NO.	LENGTH	DASH NO.
1/8	-1	1/2	-7
3/16	-2	5/8	-8
1/4	-3	3/4	-9
5/16	-4	7/8	-10
3/8	-5	1	-11
7/16	-6	1 1/4	-12

SOCKET HEAD

.140 | .078 MIN | .086 | .086 | LENGTH

CRES. PASS. MS18995

LENGTH	DASH NO.	LENGTH	DASH NO.
3/16	-1	1/2	-4
1/4	-2	5/8	
3/8	-3	3/4	

SET SCREW CUP POINT

.035 NOM | .086 | LENGTH

CRES. PASS. MS51021

LENGTH	DASH NO.	LENGTH	DASH NO.
1/8	-1	1/2	
3/16	-2	9/16	
1/4	-3	5/8	
5/16		3/4	
3/8		7/8	
7/16		1	

EXT LOCK WASHER
CRES. PASSIVATED

N/A | N/A | N/A

FLAT WASHER
CRES PASSIVATED
MS15795-802

.025 | .089 MIN | .250

INT LOCK WASHER
CRES. PASSIVATED
MS35333-80

.015 | .089 MIN | .200

SPLIT WASHER
CRES. PASSIVATED
MS35338 134

.026 | .088 MIN | .172

NUT
CRES. PASSIVATED
MS35649-224

.217 | .188 | .066

CLEARANCE HOLE

1 HOLE .089 DIA (NO. 43) | MULTIPLE HOLES

TAPPED HOLE

.070 DIA (NO. 50) 90° CSK x .086 DIA NO. 2-56 UNC-2B THD

BLIND TAPPED HOLE DRILL DEPTH

A | B

1 DIA

A	B
.09	.22

1 1/2 DIA

A	B
.13	.26

CSK HOLE

.116 DIA (NO. 32) 82° CSK x .180 DIA

C' BORE

.144 DIA (NO. 27) .312 DIA C' BORE X ____ DEEP

HELICAL HOLE - N/A

HELICAL BLIND HOLE DRILL DEPTH

1 DIA

A	B
-	-

1 1/2 DIA

A	B
-	-

2 DIA

A	B
-	-

A | B

HELICAL COIL

LENGTH	NON-LOCKING	LOCKING

LENGTH | N/A

ALL DIMENSIONS ARE MAXIMUM UNLESS OTHERWISE SPECIFIED

4·40

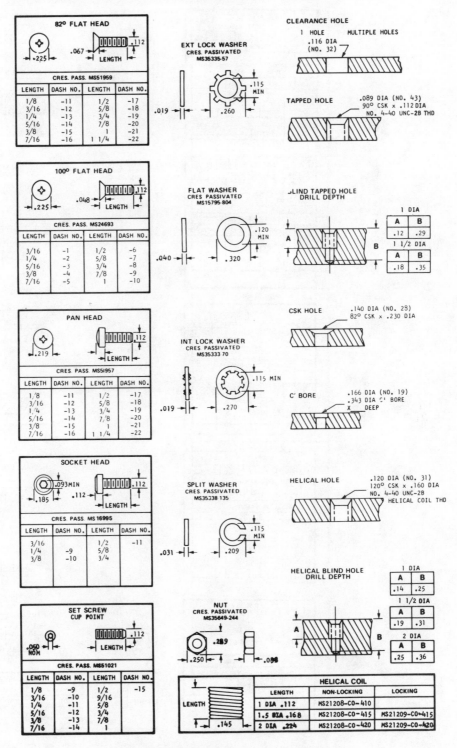

82° FLAT HEAD

.225   .067   .112   LENGTH

CRES. PASS. MS51959

LENGTH	DASH NO.	LENGTH	DASH NO.
1/8	-11	1/2	-17
3/16	-12	5/8	-18
1/4	-13	3/4	-19
5/16	-14	7/8	-20
3/8	-15	1	-21
7/16	-16	1 1/4	-22

100° FLAT HEAD

.225   .048   .112   LENGTH

CRES. PASS. MS24693

LENGTH	DASH NO.	LENGTH	DASH NO.
3/16	-1	1/2	-6
1/4	-2	5/8	-7
5/16	-3	3/4	-8
3/8	-4	7/8	-9
7/16	-5	1	-10

PAN HEAD

.219   .112   LENGTH

CRES. PASS. MS5I957

LENGTH	DASH NO.	LENGTH	DASH NO.
1/8	-11	1/2	-17
3/16	-12	5/8	-18
1/4	-13	3/4	-19
5/16	-14	7/8	-20
3/8	-15	1	-21
7/16	-16	1 1/4	-22

SOCKET HEAD

.185   .093MIN   .112   .112   LENGTH

CRES. PASS. MS16995

LENGTH	DASH NO.	LENGTH	DASH NO.
3/16		1/2	-11
1/4	-9	5/8	
3/8	-10	3/4	

SET SCREW CUP POINT

.050 NOM   .112   LENGTH

CRES. PASS. MS51021

LENGTH	DASH NO.	LENGTH	DASH NO.
1/8	-9	1/2	-15
3/16	-10	9/16	
1/4	-11	5/8	
5/16	-12	3/4	
3/8	-13	7/8	
7/16	-14	1	

EXT LOCK WASHER
CRES. PASSIVATED
MS35335-57

.019   .260   .115 MIN

FLAT WASHER
CRES PASSIVATED
MS15795-804

.040   .120 MIN   .320

INT LOCK WASHER
CRES PASSIVATED
MS35333 70

.019   .270   .115 MIN

SPLIT WASHER
CRES. PASSIVATED
MS35338 135

.031   .209   .115 MIN

NUT
CRES. PASSIVATED
MS35649-244

.219   .250   .098

CLEARANCE HOLE

1 HOLE   MULTIPLE HOLES
.116 DIA (NO. 32)

TAPPED HOLE

.089 DIA (NO. 43)
90° CSK × .112 DIA
NO. 4-40 UNC-2B THD

BLIND TAPPED HOLE DRILL DEPTH

1 DIA	
A	B
.12	.29

1 1/2 DIA	
A	B
.18	.35

CSK HOLE

.140 DIA (NO. 28)
82° CSK × .230 DIA

C' BORE

.166 DIA (NO. 19)
.343 DIA C' BORE
X   DEEP

HELICAL HOLE

.120 DIA (NO. 31)
120° CSK × .160 DIA
NO. 4-40 UNC-2B
HELICAL COIL THD

HELICAL BLIND HOLE DRILL DEPTH

1 DIA	
A	B
.14	.25

1 1/2 DIA	
A	B
.19	.31

2 DIA	
A	B
.25	.36

HELICAL COIL

LENGTH		NON-LOCKING	LOCKING
1 DIA	.112	MS21208-C0-410	
1.5 DIA	.168	MS21208-C0-415	MS21209-C0-415
2 DIA	.224	MS21208-C0-420	MS21209-C0-420

LENGTH   .145

ALL DIMENSIONS ARE MAXIMUM UNLESS OTHERWISE SPECIFIED

6·32

ALL DIMENSIONS ARE MAXIMUM UNLESS OTHERWISE SPECIFIED

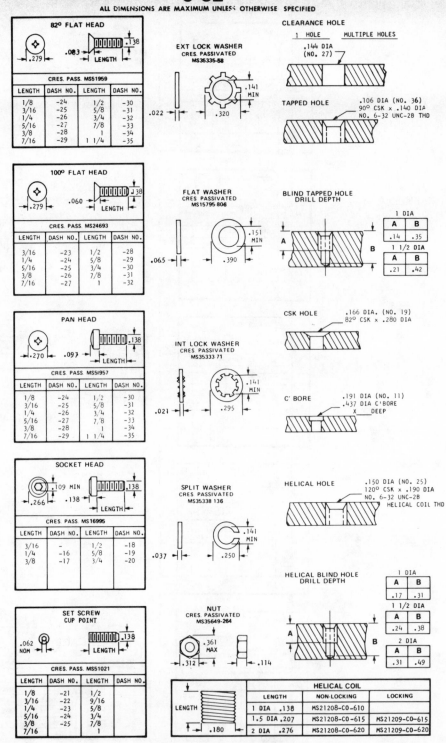

82° FLAT HEAD

.279   .083   .138   LENGTH

CRES. PASS. MS51959

LENGTH	DASH NO.	LENGTH	DASH NO.
1/8	-24	1/2	-30
3/16	-25	5/8	-31
1/4	-26	3/4	-32
5/16	-27	7/8	-33
3/8	-28	1	-34
7/16	-29	1 1/4	-35

100° FLAT HEAD

.279   .060   .138   LENGTH

CRES. PASS. MS24693

LENGTH	DASH NO.	LENGTH	DASH NO.
3/16	-23	1/2	-28
1/4	-24	5/8	-29
5/16	-25	3/4	-30
3/8	-26	7/8	-31
7/16	-27	1	-32

PAN HEAD

.270   .097   .138   LENGTH

CRES. PASS MS51957

LENGTH	DASH NO.	LENGTH	DASH NO.
1/8	-24	1/2	-30
3/16	-25	5/8	-31
1/4	-26	3/4	-32
5/16	-27	7/8	-33
3/8	-28	1	-34
7/16	-29	1 1/4	-35

SOCKET HEAD

.266   .109 MIN   .138   .138   LENGTH

CRES. PASS. MS16995

LENGTH	DASH NO.	LENGTH	DASH NO.
3/16	–	1/2	-18
1/4	-16	5/8	-19
3/8	-17	3/4	-20

SET SCREW CUP POINT

.062 NOM   .138   LENGTH

CRES. PASS. MS51021

LENGTH	DASH NO.	LENGTH	DASH NO.
1/8	-21	1/2	
3/16	-22	9/16	
1/4	-23	5/8	
5/16	-24	3/4	
3/8	-25	7/8	
7/16		1	

EXT LOCK WASHER
CRES. PASSIVATED
MS35335-88

.022   .141 MIN   .320

FLAT WASHER
CRES PASSIVATED
MS15795-806

.065   .151 MIN   .390

INT LOCK WASHER
CRES. PASSIVATED
MS35333 71

.021   .141 MIN   .295

SPLIT WASHER
CRES PASSIVATED
MS35338 136

.037   .141 MIN   .250

NUT
CRES. PASSIVATED
MS35649-264

.361 MAX   .312   .114

CLEARANCE HOLE

1 HOLE   MULTIPLE HOLES

.144 DIA (NO. 27)

TAPPED HOLE

.106 DIA (NO. 36)
90° CSK x .140 DIA
NO. 6-32 UNC-2B THD

BLIND TAPPED HOLE
DRILL DEPTH

A   B

	1 DIA
A	B
.14	.35

	1 1/2 DIA
A	B
.21	.42

CSK HOLE

.166 DIA. (NO. 19)
82° CSK x .280 DIA

C' BORE

.191 DIA (NO. 11)
.437 DIA C'BORE
X   DEEP

HELICAL HOLE

.150 DIA (NO. 25)
120° CSK x .190 DIA
NO. 6-32 UNC-2B
HELICAL COIL THD

HELICAL BLIND HOLE
DRILL DEPTH

A   B

	1 DIA
A	B
.17	.31

	1 1/2 DIA
A	B
.24	.38

	2 DIA
A	B
.31	.49

HELICAL COIL

LENGTH		NON-LOCKING	LOCKING
1 DIA	.138	MS21208-CO-610	
1.5 DIA	.207	MS21208-CO-615	MS21209-CO-615
2 DIA	.276	MS21208-CO-620	MS21209-CO-620

.180

DESIGN REFERENCE MATERIALS

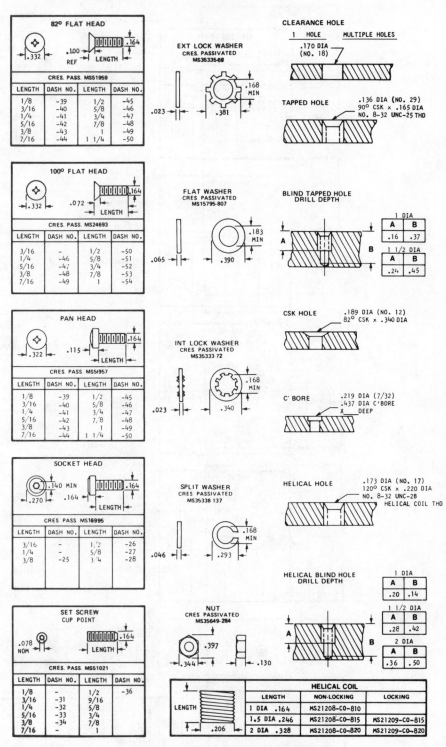

82° FLAT HEAD

.332 | .100 REF | LENGTH | .164

CRES. PASS. MS51959

LENGTH	DASH NO.	LENGTH	DASH NO.
1/8	-39	1/2	-45
3/16	-40	5/8	-46
1/4	-41	3/4	-47
5/16	-42	7/8	-48
3/8	-43	1	-49
7/16	-44	1 1/4	-50

100° FLAT HEAD

.332 | .072 | LENGTH | .164

CRES. PASS. MS24693

LENGTH	DASH NO.	LENGTH	DASH NO.
3/16	–	1/2	-50
1/4	-46	5/8	-51
5/16	-47	3/4	-52
3/8	-48	7/8	-53
7/16	-49	1	-54

PAN HEAD

.322 | .115 | LENGTH | .164

CRES. PASS. MS51957

LENGTH	DASH NO.	LENGTH	DASH NO.
1/8	-39	1/2	-45
3/16	-40	5/8	-46
1/4	-41	3/4	-47
5/16	-42	7/8	-48
3/8	-43	1	-49
7/16	-44	1 1/4	-50

SOCKET HEAD

.140 MIN | .270 | .164 | LENGTH | .164

CRES. PASS. MS16995

LENGTH	DASH NO.	LENGTH	DASH NO.
3/16	–	1/2	-26
1/4	-25	5/8	-27
3/8		3/4	-28

SET SCREW
CUP POINT

.078 NOM | LENGTH | .164

CRES. PASS. MS51021

LENGTH	DASH NO.	LENGTH	DASH NO.
1/8	–	1/2	
3/16	-31	9/16	-36
1/4	-32	5/8	
5/16	-33	3/4	
3/8	-34	7/8	
7/16	–	1	

EXT LOCK WASHER
CRES. PASSIVATED
MS35335-60

.023 | .168 MIN | .381

FLAT WASHER
CRES PASSIVATED
MS15795-807

.065 | .183 MIN | .390

INT LOCK WASHER
CRES PASSIVATED
MS35333-72

.023 | .168 MIN | .340

SPLIT WASHER
CRES PASSIVATED
MS35338-137

.046 | .168 MIN | .293

NUT
CRES PASSIVATED
MS35649-284

.397 | .344 | .130

CLEARANCE HOLE

| 1 HOLE | MULTIPLE HOLES |

.170 DIA (NO. 18)

TAPPED HOLE

.136 DIA (NO. 29)
90° CSK × .165 DIA
NO. 8-32 UNC-25 THD

BLIND TAPPED HOLE
DRILL DEPTH

1 DIA	
A	B
.16	.37

1 1/2 DIA	
A	B
.24	.45

CSK HOLE

.189 DIA (NO. 12)
82° CSK × .340 DIA

C'BORE

.219 DIA (7/32)
.437 DIA C'BORE
X DEEP

HELICAL HOLE

.173 DIA (NO. 17)
120° CSK × .220 DIA
NO. 8-32 UNC-2B
HELICAL COIL THD

HELICAL BLIND HOLE
DRILL DEPTH

1 DIA	
A	B
.20	.14

1 1/2 DIA	
A	B
.28	.42

2 DIA	
A	B
.36	.50

HELICAL COIL

LENGTH | .206

LENGTH	NON-LOCKING	LOCKING
1 DIA .164	MS21208-C0-810	
1.5 DIA .246	MS21208-C0-815	MS21209-C0-815
2 DIA .328	MS21208-C0-820	MS21209-C0-820

ALL DIMENSIONS ARE MAXIMUM UNLESS OTHERWISE SPECIFIED

DESIGN REFERENCE MATERIALS

10·32

ALL DIMENSIONS ARE MAXIMUM UNLESS OTHERWISE SPECIFIED

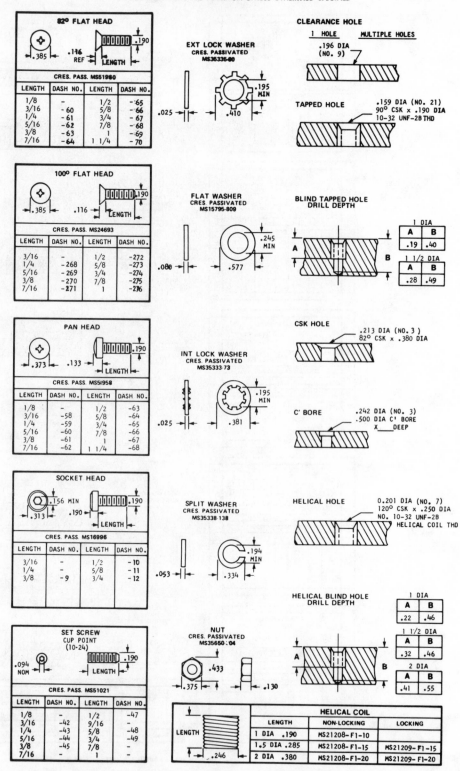

82° FLAT HEAD

.385 .116 REF .190 LENGTH

CRES. PASS. MS51960

LENGTH	DASH NO.	LENGTH	DASH NO.
1/8	–	1/2	-65
3/16	-60	5/8	-66
1/4	-61	3/4	-67
5/16	-62	7/8	-68
3/8	-63	1	-69
7/16	-64	1 1/4	-70

100° FLAT HEAD

.385 .116 .190 LENGTH

CRES. PASS. MS24693

LENGTH	DASH NO.	LENGTH	DASH NO.
3/16	–	1/2	-272
1/4	-268	5/8	-273
5/16	-269	3/4	-274
3/8	-270	7/8	-275
7/16	-271	1	-276

PAN HEAD

.373 .133 .190 LENGTH

CRES. PASS. MS51958

LENGTH	DASH NO.	LENGTH	DASH NO.
1/8	–	1/2	-63
3/16	-58	5/8	-64
1/4	-59	3/4	-65
5/16	-60	7/8	-66
3/8	-61	1	-67
7/16	-62	1 1/4	-68

SOCKET HEAD

.156 MIN .313 .190 .190 LENGTH

CRES. PASS. MS16996

LENGTH	DASH NO.	LENGTH	DASH NO.
3/16	–	1/2	-10
1/4	–	5/8	-11
3/8	-9	3/4	-12

SET SCREW
CUP POINT
(10-24)

.094 NOM .190 LENGTH

CRES. PASS. MS51021

LENGTH	DASH NO.	LENGTH	DASH NO.
1/8	–	1/2	-47
3/16	-42	9/16	–
1/4	-43	5/8	-48
5/16	-44	3/4	-49
3/8	-45	7/8	–
7/16	–	1	–

EXT LOCK WASHER
CRES. PASSIVATED
MS35335-80

.025 .195 MIN .410

FLAT WASHER
CRES. PASSIVATED
MS15795-809

.080 .245 MIN .577

INT LOCK WASHER
CRES. PASSIVATED
MS35333-73

.025 .195 MIN .381

SPLIT WASHER
CRES PASSIVATED
MS35338-138

.053 .194 MIN .334

NUT
CRES. PASSIVATED
MS35650-04

.433 .375 .130

HELICAL COIL

LENGTH .246

	LENGTH	NON-LOCKING	LOCKING
1 DIA	.190	MS21208-F1-10	
1.5 DIA	.285	MS21208-F1-15	MS21209-F1-15
2 DIA	.380	MS21208-F1-20	MS21209-F1-20

CLEARANCE HOLE

1 HOLE   MULTIPLE HOLES

.196 DIA (NO. 9)

TAPPED HOLE

.159 DIA (NO. 21)
90° CSK × .190 DIA
10-32 UNF-2B THD

BLIND TAPPED HOLE
DRILL DEPTH

1 DIA	
A	B
.19	.40

1 1/2 DIA	
A	B
.28	.49

CSK HOLE

.213 DIA (NO. 3)
82° CSK × .380 DIA

C' BORE

.242 DIA (NO. 3)
.500 DIA C' BORE
X____DEEP

HELICAL HOLE

0.201 DIA (NO. 7)
120° CSK × .250 DIA
NO. 10-32 UNF-2B
HELICAL COIL THD

HELICAL BLIND HOLE
DRILL DEPTH

1 DIA	
A	B
.22	.46

1 1/2 DIA	
A	B
.32	.46

2 DIA	
A	B
.41	.55

DESIGN REFERENCE MATERIALS

1/4·28

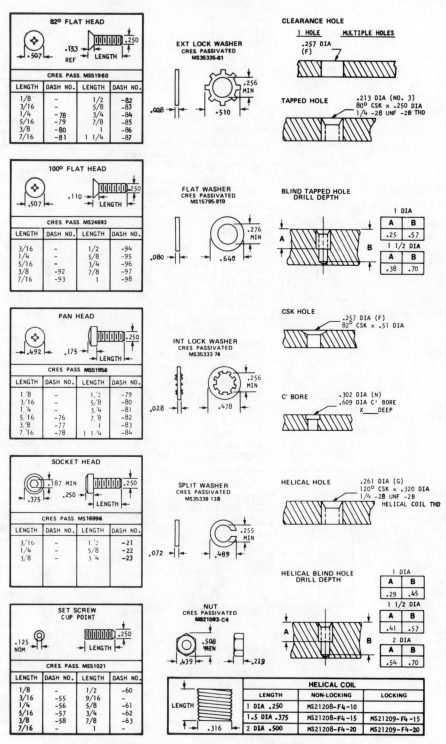

82° FLAT HEAD

.507 | .153 REF | LENGTH | .250

CRES. PASS. MS51960

LENGTH	DASH NO.	LENGTH	DASH NO.
1/8	–	1/2	-82
3/16	–	5/8	-83
1/4	-78	3/4	-84
5/16	-79	7/8	-85
3/8	-80	1	-86
7/16	-81	1 1/4	-87

100° FLAT HEAD

.507 | .110 | LENGTH | .250

CRES. PASS. MS24693

LENGTH	DASH NO.	LENGTH	DASH NO.
3/16	–	1/2	-94
1/4	–	5/8	-95
5/16	–	3/4	-96
3/8	-92	7/8	-97
7/16	-93	1	-98

PAN HEAD

.492 | .175 | LENGTH | .250

CRES. PASS. MS51958

LENGTH	DASH NO.	LENGTH	DASH NO.
1/8	–	1/2	-79
3/16	–	5/8	-80
1/4	–	3/4	-81
5/16	-76	7/8	-82
3/8	-77	1	-83
7/16	-78	1 1/4	-84

SOCKET HEAD

.375 | .187 MIN | .250 | LENGTH | .250

CRES. PASS. MS16996

LENGTH	DASH NO.	LENGTH	DASH NO.
3/16	–	1/2	-21
1/4	–	5/8	-22
3/8	–	3/4	-23

SET SCREW CUP POINT

.125 NOM | LENGTH | .250

CRES. PASS. MS51021

LENGTH	DASH NO.	LENGTH	DASH NO.
1/8	–	1/2	-60
3/16	-55	9/16	–
1/4	-56	5/8	-61
5/16	-57	3/4	-62
3/8	-58	7/8	-63
7/16	–	1	–

EXT LOCK WASHER
CRES. PASSIVATED
MS35335-61

.028 | .256 MIN | .510

FLAT WASHER
CRES PASSIVATED
MS15795-810

.080 | .276 MIN | .640

INT LOCK WASHER
CRES. PASSIVATED
MS35333-74

.028 | .256 MIN | .478

SPLIT WASHER
CRES PASSIVATED
MS35338 138

.072 | .255 MIN | .489

NUT
CRES. PASSIVATED
MS21083-C4

.508 MIN | .439 | .219

CLEARANCE HOLE

1 HOLE MULTIPLE HOLES
.257 DIA (F)

TAPPED HOLE

.213 DIA (NO. 3)
80° CSK x .250 DIA
1/4 -28 UNF -2B THD

BLIND TAPPED HOLE DRILL DEPTH

1 DIA	
A	B
.25	.57

1 1/2 DIA	
A	B
.38	.70

CSK HOLE

.257 DIA (F)
82° CSK x .51 DIA

C' BORE

.302 DIA (N)
.609 DIA C' BORE
X___ DEEP

HELICAL HOLE

.261 DIA (G)
120° CSK x .320 DIA
1/4 -28 UNF -2B
HELICAL COIL THD

HELICAL BLIND HOLE DRILL DEPTH

1 DIA	
A	B
.29	.45

1 1/2 DIA	
A	B
.41	.57

2 DIA	
A	B
.54	.70

HELICAL COIL

LENGTH	NON-LOCKING	LOCKING
1 DIA .250	MS21208-F4-10	
1.5 DIA .375	MS21208-F4-15	MS21209-F4-15
2 DIA .500	MS21208-F4-20	MS21209-F4-20

LENGTH | .316

ALL DIMENSIONS ARE MAXIMUM UNLESS OTHERWISE SPECIFIED

DESIGN REFERENCE MATERIALS

INDEX

INDEX

389